THE
JAPANESE GOVERNMENT
1/4 RUPEE
THE
JAPANESE GOVERNMENT
M/AD
ONE CENT
M/AD
THE JAPANE
FIVE
ML
JAPA
M/BE
MM
THE
JAPANESE GOVERNMENT
TEN
CENTS
MX
MX
MT
THE
JAPANESE GOVERNMENT
PROMISES TO PAY THE BEARER ON DEMAND
FIFTY CENTS
MT
50
50
50
50

The Survival Factor

HOKKAIDO
Hakodate
JAPAN
Sakata
Sendai
JAPAN SEA
Tokyo
Yokohama
KOREA
HONSHU
Keijo
Osaka
YELLOW SEA
SHIKOKU
Moji
KYUSHU
Nagasaki
NORTH
EAST CHINA SEA
Shanghai
Nanking
RYUKYO
Okinawa
ARCH^O
PACIFIC
CHINA
Taihoku
TAIWAN (Formosa)
Yangtze
Chungking
Amoy
Tainan
OCEAN
ROAD
Canton
Hong Kong
LUZON
PHILIPPINE Is
Salween
Kunming
BURMA
Kiungchow
Manila
HAINAN
Hanoi
Lashio
BURMA
Mekong
FRENCH INDO-CHINA
MINDANAO
Mandalay
PALAWAN
THAILAND
SOUTH CHINA SEA
Rangoon
Moulmein
Bangkok
Saigon
N^TH BORNEO
Ye
BAY OF BENGAL
Tavoy
Mergui
Phnom Penh
SARAWAK
Andawan Is
CELEBES
MALAYA
BORNEO
Nicobar Is
Medan
Singapore
SUMATRA
C.H. Richards

The Survival Factor

Rowley Richards
and Marcia McEwan

Costello

This book is dedicated to Jim Armstrong, all volunteer medical orderlies, their patients and their relatives.

First published in Great Britain by D J Costello (Publishers) Ltd
43 High Street Tunbridge Wells Kent TN1 1XL
Printed in Australia by Southwood Press Pty Ltd

ISBN 0 7104 2066 8

Contents

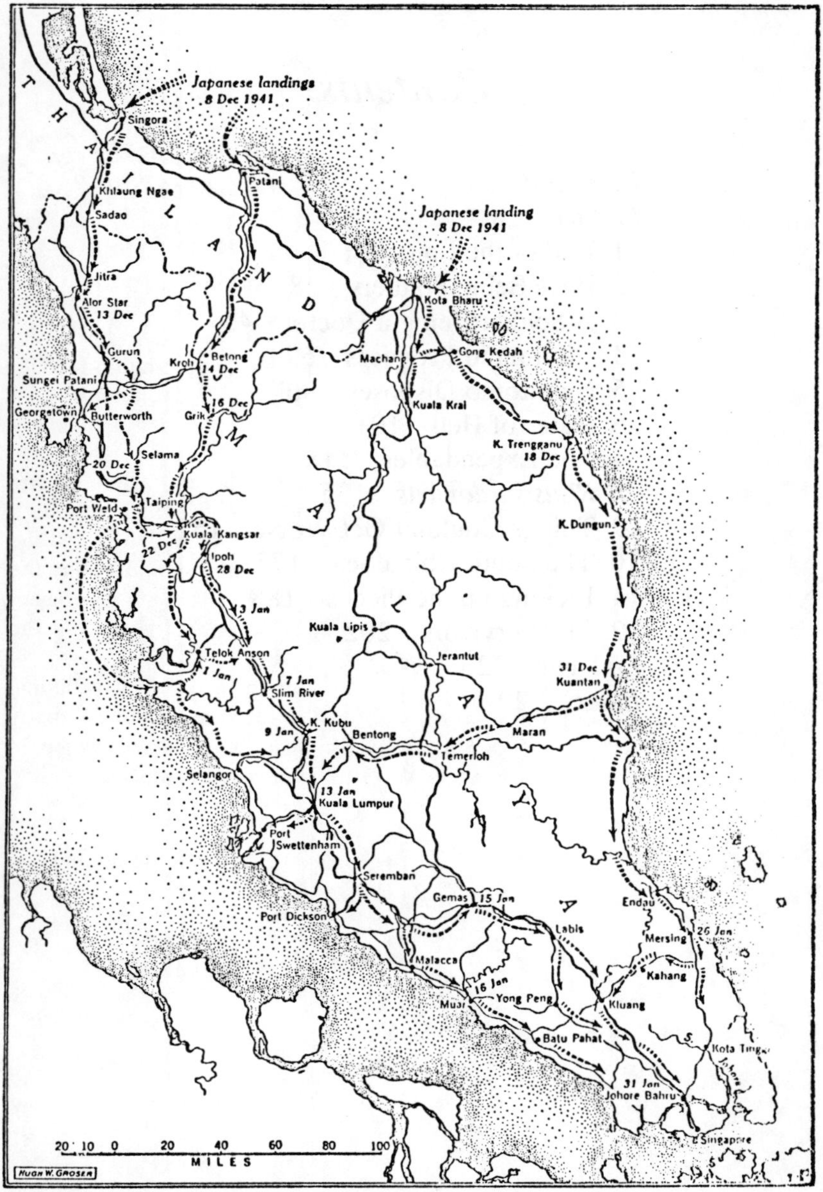

The conquest of Malaya

Foreword

This is a story which will shock, upset, uplift and perhaps inspire the reader—whether the reader was a prisoner of war, like Rowley Richards, or a civilian who knows little about the Malayan Campaign of 1941–42 and the period of three years and six months spent as a prisoner of war of the Japanese.

The book will shock many people because it has the ring of authenticity and describes in simple language the actual horrors of brutality, sadism and death suffered by the prisoners. It will upset the reader that these things could happen and did happen in a world he thought was civilised, and unless we are vigilant, could happen again. It will uplift the reader who may remember Shakespeare's immortal words in King Henry V—'We few, we happy few, we band of brothers'—for as the author points out again and again, no one could have survived without his mates who became closer than his brothers.

For those who survived the aftermath of tortures, the degradation and the crushing medical problems inflicted upon them, their story should inspire the reader. For the ex-prisoner, every day is a bonus and time must be spent considering and actively working for the welfare of one's friends and the community at large.

Perhaps Dr Richards should add another factor to his survival equation. His own courage and devotion to the sick men in his care resulted in many survivors—I know—because he saved my life on at least two occasions in Burma, and I am here, privileged to write this Foreword to his outstanding book.

Roy H. Whitecross

Preface

The sudden declaration of war, after a generation of peace, uproots young men, many of them barely out of school. Security is gone, plans for the future are set aside: they are fed to the military machine to be transformed into soldiers, taught to hate, taught to kill automatically, trained to forget the Christian ethic, the lifetime teaching of 'Thou shalt not kill'. It is harsh. But it is the survival mechanism of war even for those who are the non-aggressors and have had war thrust on them. It has to be done.

But what is the survival mechanism of the defeated, the prisoner of war? How are soldiers who have been trained to fight, but are then deprived of their raison d'être, to cope with the realities of defeat? While the survival of soldiers at war depends on the interaction of numerous factors, the survival of POWs depends largely on three factors—the treatment meted out by their captors; the conditions under which they live; and the devotion, loyalty, initiative and comradeship of their fellow prisoners. In the war in the Pacific, 1941–45, Allied prisoners of war were treated like criminals; used as slave labour; subjected to the rotting hell of jungle that is inimical to human life, particularly to white men; killed by tropical diseases; starved.

The Survival Factor explores the medical problem of keeping prisoners of war alive under horrifying conditions through disaster after crushing disaster—from the building of the Burma railway, through shipwreck, to the last bitter winter of the war, in Japan. It is a medical story told by a regimental medical officer who was with his men all the way.

It is fact not fiction. It is a history but not dead history. It is pertinent for today's young men and women who ask, at a time of crisis, 'What is war really like?' It is the history of men who survived when all the odds were against them. It seeks an answer to the question: 'What is the unknown quantity—the '*s*' factor in the survival equation that was common to those few who did survive?'

This story of survival of some prisoners of war of the Japanese addresses this question and in doing so highlights the dedication of previously untrained medical orderlies and is based on personal diaries, official documents and post-war publications.

This book is based on my recollections of events as one individual perceived and recorded them at the time.

Three criteria were observed in writing this work. The information must be factual, the dialogue must be feasible, and both must be credible to those readers who did not share the experience. All characters are real, with the exception of the 'Dodger' and his mates whose composite identities shall remain anonymous.

Many individuals have contributed to the work.

Sergeant Jim Armstrong and his team of medical orderlies together with their patients and all the soldiers with whom I served provided the inspiration. Writing the manuscript was realised with the encouragement of my family and other friends.

Information to supplement and complement my personal diary and records was provided from documents and records kept by my medical orderlies and assistants, and by the personal diaries of Sergeant Jim Lynch (loaned by his sister, Mrs K. Harrison); Sergeant Scippy Maher (loaned by his widow, Mrs Joan Maher); Lieutenant-Colonel John Williams; and Mr Roy Whitecross, the author of *Slaves of the Son of Heaven.* Additional information was obtained from the following: the Australian War Memorial; the National Archives of the United States; U.S. Navy Archives (Mr D. Cavalcante); *The Japanese Thrust* by Lionel Wigmore; *Submarine* by Theodore Roscoe; *Return from the River Kwai* by Joan and Clay Blair; and the Japanese Embassy, Canberra (Captain A. Terai).

The original manuscript was typed in 1946 by Mrs Mary Leggo (nee Best) and checked by Major John Shaw. The final manuscript was typed by Mrs Anne Stohr, and checked by my wife, Beth.

To each and every one of them and many others too numerous to mention, I offer my sincere gratitude for their encouragement, persistence and constructive criticism.

For editorial and other advice in the final stages of preparation I thank Carl Harrison-Ford, John Brennan and Ron Brennan. Thanks also to Douglas Albion, for 25 years an executive with *Reader's Digest* in Australia, who introduced me to my joint author, Marcia McEwan, to whom I pay particular tribute for the sympathetic, perceptive and realistic manner in which she has brought the words to life.

Rowley Richards,
January 1989

BURMA

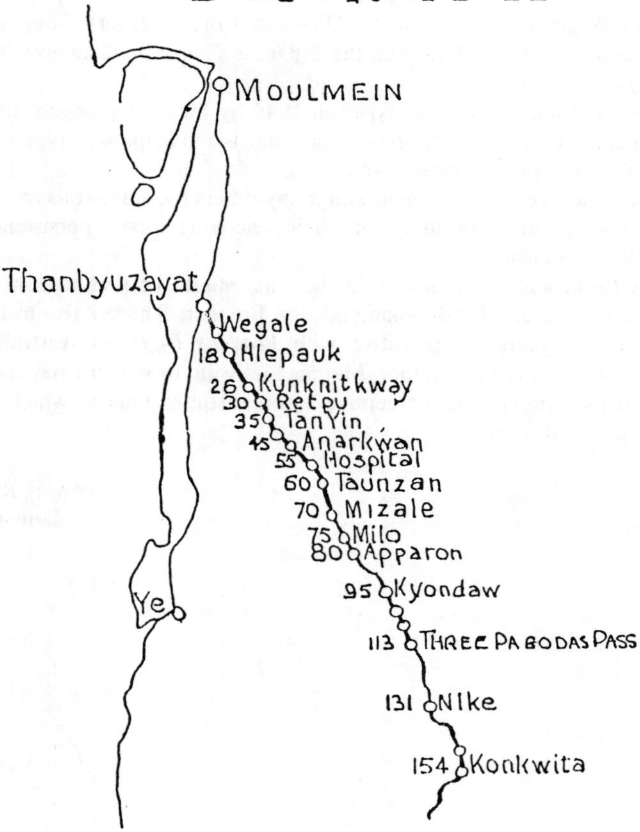
MOULMEIN
Thanbyuzayat
Wegale
18 Hlepauk
26 Kunknitkway
30 Retpu
35 TanYin
45 Anarkwan
55 Hospital
60 Taunzan
70 Mizale
75 Milo
80 Apparon
95 Kyondaw
113 THREE PAGODAS PASS
131 Nike
154 Konkwita
Ye

1

End of the Line

On the morning of 17 October 1943 the railway construction camp at the 131-kilometre peg from Thanbyuzayat was quiet.

The two sections of the Burma-Thailand railway were due to be joined at the 154-kilometre peg, near Three Pagodas Pass, on that day. Joining Moulmein in Burma with Bangkok in Thailand, the railway had been built on the initiative of the Japanese but constructed with prisoner-of-war labour, through 270 miles of some of the worst jungle in the world. For the Japanese it was a big celebration day, but for the prisoners in the jungle construction camps it was just another day of being sick and hungry and wondering what bloody-minded job the Japs would have for them next. British, Australian, American and Dutch prisoners of war had died in thousands, of which the final tally had not been made, and used up as callously as ever the Pharoahs of Egypt had used their slaves. For the survivors their sufferings were not yet over. They endured doggedly, one day at a time, with the hope that each day brought them closer to freedom. Without that hope they could not have endured.

The prisoners of No. 3 Branch Thai POW Camp occupying the 131-kilo camp or Little Nike, as it was known to them, were Australians—'Anderson' and 'Williams' Forces. Every fit man, other than the cooks and the medical staff, had been sent out to work on the line. The No Duties sick rested on the bamboo platforms that were their beds in the long bamboo and atap sleeping hut. At the end of the hut was a section designated as 'Regimental Aid Post' (RAP) and 'hospital', where the very sick were attended by two doctors and two medical orderlies. Outside the hut light rain was falling, although this was the beginning of the Dry season, turning the bare earth of the 'parade ground' into mud again. Beyond the parade ground was the other enemy, the jungle, rotting and flourishing in rottenness, radiating a smell of decay that was almost tangible, almost audible, breeding disease. The Regimental Medical Officer (known as the RMO) hated both enemies, the Nips and the jungle, and in hatred had found the strength and cunning to endure and do battle with them.

He did not care that the line would be joined that day. He cared that Gunner Chauncey, a game little battler who had not given up the battle yet, would probably die; that Sergeant Jim Armstrong, his medical orderly, was one of the hospitalised, thin and exhausted with diarrhoea; that Doctor John Chalmers' face was an ashen mask of strain out of which looked bewildered, suffering eyes. Chalmers, bending over Corporal Bob Kemp, the orderly with beriberi, was a good doctor and a brave man, but a 'hospital' doctor who should not be in the jungle.

They should have known better at the headquarters hospital than to send him out here, the RMO thought angrily. There's no way he could cope, being thrown in to it like this. At HQ they have beds with sheets, drugs—not many but some—even bedpans and a canteen. We have the jungle and our wits.

Turning away from his patient, Major John Chalmers looked at the RMO. Why the ferocious scowl, he wondered. There hasn't been a major dust-up since morning sick parade, but Rowley doesn't look his sunny self.

Captain Rowland Richards, Regimental Medical Officer, 2/15 Australian Field Regiment, Medical Officer for the Japanese POW No.2 Battalion of 'A' Force, known as 'Anderson' Force, had recently celebrated his twenty-seventh birthday although physically he appeared much younger. He wasn't tall—just five foot seven—and had always been lightly built. Now he was underweight. His neat-featured, fresh-skinned face and ready smile that quirked up to the left had earned him the nickname of 'Baby Doctor' in the training camps at Ingleburn and Holdsworthy near Sydney. But that was three years ago, before any of the young men, most younger than their RMO, knew what war was really about.

Now his fresh complexion was burned red by the tropical sun and his smile had set into an almost-permanent sardonic twist of the lips. He was alert and quick-speaking, almost terse: an old young man. God, what has it really been like out here, Chalmers wondered. I've only had a few weeks and it's hell. How do they stand it?

'At least it's quiet outside,' he said.

'Don't bet on it staying that way,' Rowley answered. 'The Nips'll find something to scream about.'

As if to prove the RMO clairvoyant the shouting of high-pitched voices shattered the silence and the rubber boots of running Korean guards slapped and sucked in the mud. There was a sharp crack of a bamboo pole smacked against the bambo uprights of the opening to the sleeping hut. A shrill, rapid voice demanded instant obedience.

'All men, *shigoto, speedo, speedo*.'

That sounded like 'The Jeep' wanting men for a work party—and in a hurry.

A few Australian voices replied in groaned or muttered complaint. 'Jesus!' someone with more strength or hardihood sang out. 'What do the little bastards want now?'

Chalmers tensed. The rising heart rate of nervous tachycardia made him feel as if his whole chest was vibrating, and his throat constricted.

Sergeant Jim Armstrong relaxed his effort to get out of his cot to use the tin hat that was a substitute for the enamel bedpan of a proper hospital. He leaned against the supporting arm of the red-headed orderly, Corporal Pinky

Rhodes. One of the most experienced troop medical orderlies, Rhodes was usually at work on the railway so immediate attention would be available if a worker was injured or taken ill. But today Rowley had noticed signs of exhaustion and kept him in camp.

'Maybe the embankment's slipped into the gully, line and all,' Pinky said hopefully.

The young soldier who had been drafted by Rowley out of the Light Duties sick to assist the medical orderlies hunched his shoulders and concentrated all his attention on placing a dressing over the ugly shin sore of one of the tropical ulcer cases, exercising his carefully developed talent for making himself inconspicuous when a row was brewing. He was glad he was in the RAP. You needed to know how to be out of sight out of mind when the Monkeys were carrying on.

Men were rolling off their sleeping platforms, stumbling out to assemble in the mud, and Rowley followed them briskly, carrying his slight frame very straight. He worked hard to maintain that military bearing, even when every bone ached with fatigue. It commanded some respect from the guards without infuriating them, as it would have done had he been a few inches taller. They hated having to look up at a prisoner.

Rowley squinted his eyelids against the rain, which had been coming in intermittent showers since before dawn when the work party had been herded out of the camp.

What did the Nips want now? It looked as if they were assembling another work party.

That wasn't on. In the pre-dawn darkness, lighted by fires of heaped bamboo and occasional lanterns, the Nips had ignored his bargaining and taken every man capable of standing upright and putting one foot in front of the other. The fittest of them had already worked an unbroken thirty-three hour shift, and another of eighteen hours before that. The convalescent men marked as Light Duties had been taken as well. Rowley had not pressed his arguments on their behalf too hard. He knew when the guards were not going to give any ground and he also knew that this was not going to be a heavy day's work. The railway had been virtually completed during the last shift. There was only tidying up to be done. So the Light Duties sick would not be harmed by a one-day shift on the line. Rowley was satisfied that the No Duties men had been left. Why, then, were the bloody bastards hunting the No Duties men out of their quarters and lining them up?

As briskly as he could manage in his homemade clogs—thick wooden platforms with a strap of hide to hold the foot that had been his usual footware since his boots began to wear—he approached the Korean guard. He saluted. That was required, even though he was an officer and the Jeep was a third-class private and a Korean as well; bottom of the hierarchy in the Imperial Japanese Army. Prisoners were even lower than third-class Korean privates.

Rowley even bowed slightly since he was approaching the Jeep outside the official routine of the sick parade. It was politic to observe ritual politeness; not bowing, looking a Nip straight in the eyes, was impolite, even insulting.

Perfunctorily the Jeep returned the bow with a slightly lesser degree of inclination although he looked as if he would rather have dealt Rowley a clout over the ear. He started to turn away, thumping the ground with his bamboo stick and shouting '*Shigoto, speedo*' equally, it seemed, to the prisoners and to his own men.

'No!' Rowley said firmly and loudly. '*All byoki. No shigoto.*' (All sick. No work.)

'*Shigoto.*' the Jeep sounded adamant that the men were to work. He hesitated, then unexpectedly offered an explanation. 'Special for Japanese celebration. Engineers order. Little bit *shigoto.*'

He whacked his stick in the mud a fraction of an inch from Rowley's toes, due warning that the next blow would be aimed at shoulder or head if the captain doctor persisted in being a nuisance.

'*Asoka*? [Is that so?]' Rowley shrugged and retreated to the hut doorway.

'What does he want them for?' Chalmers asked.

'Special for Japanese celebration work,' Rowley mimicked.

'What?'

'Heaven only knows.'

Chalmers looked bewildered as he contemplated the scene in front of the prisoners' quarters. About a dozen men were being cuffed into a straggling line. They were ragged. Some had shirts but were bare-footed; others had clogs or the wired-together remains of rotting boots, but no shirts. Some wore only G-strings. Most of them slumped in the middle, chronic diarrhoea of course, and their bellies were the bloated pots of malnutrition. A man at the rear of the line was scratching his left ankle with the toes of his right foot. The irritation there would develop into a tropical ulcer in a couple of days if the skin was broken.

The rain had turned off again. The tail end of the unexpected work party, with its escort of yelling Korean guards, was disappearing into the gloom of the jungle. Rowley glanced at Chalmers with almost a glint of amusement in his eyes. Amusement? He was an odd young man. Here he was adopting an offhand attitude to this incident, and yet at the dawn sick parade he had argued the toss with the commander of the guards in order to keep every one of those men in camp.

Certainly he was not exactly orthodox. Chalmers remembered that there had been rumours of friction higher-up where Rowley Richards was concerned; nothing specific of course, more in what was left unsaid than what was said about a clash of opinions.

'Why did you let them go?' Chalmers asked as they went back to the hospital area.

'Why not, John?'

'But some of those chaps could hardly walk for colic or fever.'

'And for some of them it was just an unofficial rest day. You know from Jim and Allen Taylor how we work the sick parade. We try to conserve the strength of the men by giving everyone regular unofficial rest days, even if it means sending out the mildly sick who have been off duty so that men who have been working long shifts can get a break. I have the very sick safely in hospital. Those with colic can manage a short chore . . . whatever it is . . . without coming to much harm.'

'The sick aren't safe anywhere,' Chalmers said bitterly. 'If the Nips want them they take them.'

Chalmers had become increasingly horrified by conditions during his short time at Little Nike. The way the Nips disregarded the uncertain sanctuary of the RAP hospital was, to him, another appalling example of the Japanese total disregard for the international conventions governing the treatment of prisoners of war. During Rowley's absence he had stood by helpless while they blitzed the hospital and took sick men to work.

The RAP hospital was not even a hospital as understood in the permanent hospital camps. It was the regimental aid post of a line constuction camp: a typical jungle hut, long and narrow with a high-pitched roof. The frame and rafters were of six-inch diameter bamboo, roof and walls were of the usual palm-leaf thatch called atap. The atap was adequate protection during the dry months (there were only two seasons in the jungle, the Wet and the Dry), and against the intermittent tropical showers that turned off and on every half hour or so during the early weeks of the Dry. But it was not proof against the constant downpour, the driving, blinding sheets of water that was the Wet. Fortunately, however, the 131-kilo camp was newly constructed and the Wet was almost over. The cleared parade area outside the huts was again a mud rink after several days of showers, the dirt floor of the RAP was dry and swept clean. Because the huts were new there were no grubs in the atap, no bed bugs, yet, in the bamboo.

Raised platforms of bamboo slats were built down either side of the hut and divided into sleeping bays. This was standard design for prisoners' quarters in the construction camps. The bamboo platforms were both living space and bed and a man did not have to be as delicate as the princess in 'The Princess and the Pea' to feel the bamboo corrugations, especially when the only bedding was a single blanket. Necessity, ingenuity, and the talent for improvisation that had furnished many a bush hut in the pioneering days of Australia had devised the Burma Railway cot: a simple frame of bamboo supporting two bamboo poles run through either side of large rice sacks.

The medical orderlies had hoarded rice sacks for the RAP cots as they moved from camp to camp. Bamboo was the one plentiful commodity in the lives of the prisoners, and an integral part of the wall that imprisoned them more surely than the wire of Tavoy or the concrete of Changi. The wall of jungle was dark and dank, towering hugely along either side of the narrow twilit clearing through which the line was being built. Bamboo was all the jungle gave them.

At some camps there had been separate RAP huts or tents, but at Little Nike the RAP was housed in one end of a sleeping hut. Within this shelter of bamboo the sick men lay on their cots. There were eight sleeping spaces to a bay in the RAP, allowing two feet of shoulder-room per patient. The tropical ulcers and vitamin deficiency cases, beriberi and pellagra were separated from the dysentery and malaria. The man with cerebral malaria was in a bay of his own. Thanks to the scrounging talents of the orderlies, some of the sickest were under mosquito nets. A few nets were fairly recent issue, but most had been preserved for months under the loving supervision of Jim Armstrong. Those rent by hallucinating second-stage pellagra victims had been patched, others that had covered dying cholera cases had been sterilised in boiling water and were in use again.

At the far end of the hut makeshift shelves formed the RAP 'dispensary'. Most of the equipment there was also makeshift after so long without any proper issue of supplies. The few precious drugs and syringes were in a metal chest, now doing double duty as a seat for the RMO's trestle table.

The miserable inadequacy of the whole set-up had caused a permanent knot in the pit of John Chalmers' stomach during the six weeks he had been in charge. He had been sent out when Rowley was returned to Retpu hospital camp for a rest. Although news drifted back to Retpu that the construction camps up the line were hells, only experience hammered home the truth of it. Conditions offended against all Chalmers held sacred as a doctor.

The camp commander and the guards were impossible. They wanted men to work, sick or not, and there was no way of reasoning with the yelling, bashing animals. They used fists, even rifle butts, to win any argument. Chalmers had no doubt that they would, if thwarted, ram the business end of a rifle into an offender's stomach and pull the trigger. The opinion of the prisoners' medical officers meant little.

A surgeon in civilian life, Chalmers had volunteered at the outbreak of war but was a doctor rather than a soldier, despite the uniform. He believed in the tradition of professional courtesy and cooperation among medical men that transcended the artificial enmity of war. The attitude of the Japanese doctors at the POW hospital camps had generally done no great violence to these ideals. There was a certain professional courtesy even if they continually maintained that they were unable to supply drugs and equipment in sufficient quantities. But camps where filthy, drunken, illiterate peasants in uniform ruled on the medical condition of prisoners and shouted that qualified doctors were liars had stunned Chalmers. He had been stunned physically too—clouted over the head more than once when he had tried to argue.

Chalmers knew that some of the men thought he was scared stiff of the Nips. That wasn't so. His courage had been tested in civilian life and in war. As a doctor he would stand up for his medical opinions against anyone, and under fire during the confused two months that was all there was to the defence of the Malay peninsula and Singapore, he had carried out his duties without fear or hesitation. But this jungle business wasn't any sort of war that he had ever read of or heard about.

It wasn't a fighting war. For the Australians, British, Indians and Dutch who had been thrown against the southward thrust of the Japanese down the archipelago and into the islands of the East Indies, the fighting war was over.

They were the conquered, the chess pieces taken and set aside in the game of war. Yet they were dying by the hundreds, probably thousands, fighting a war for survival against both their captors and the jungle and its diseases.

How did you fight brutal captors and an environment inimical to man? John Chalmers had no ready answer, and neither of the commanding officers of the Australian forces in the jungle could give him one. Lieutenant-Colonel Anderson had one method; Lieutenant-Colonel Williams another. Brigadier Varley at headquarters was waging a war of tactics against Colonel Nagatomo, seeking concessions for his men. Major Ted Fisher was running a base hospital at Retpu with fairly decent equipment and couldn't imagine what the practice of medicine in a construction camp was really like. So Chalmers suffered because he could not do the best for the men in his care. Constantly bewildered, he wondered if the youthful Captain Rowland Richards, who grinned and was damnably offhand, had discovered the way to fight Nips and jungle and survive under these dreadful conditions. Certainly the men appeared to have faith in him.

Rowley did not have an answer. He had not found a Philosopher's Stone. But he had theories, a passion for solving problems, and tenacity. He had mentally reduced the whole show into a problem like the maths problems that had been his delight and hobby since mathematics was first introduced into his schoolboy life. And painstakingly, he worked at the problems step by step. At least he knew the nature of the problem. So did every man on the line: it was, simply, how to stay alive. In his equation 'S' equalled Survival; 'a' equalled the Nips and their disregard for human life which was expendable in their pursuit of an objective; 'b' equalled the jungle, disease and starvation; and 'c' equalled catastrophe such as acts of war. With 'a' plus 'b' plus 'c' no one should have survived. Yet some had died despite apparent physical fitness and the will to live. Others had survived against all probability. What, in human beings, was the unknown quantity: not 'x', but '*s*', the survival factor? He saw the formula $S = \mathbf{s} / (a + b + c)$.

He might never know the answer to '*s*' in his lifetime, whether his life extended into old age or ended in the jungle next week, or next month, but on 17 October 1943 he was occupied with one tiny segment of the problem. He was a doctor trying to be a doctor without equipment and trying to keep the men in the RAP hospital of 'Anderson' Force at the 131-kilo camp alive.

Jim Armstrong had settled back on his cot, relieved for the next hour or so of the wrenching abdominal cramps. Pinky Rhodes had tucked the blanket around him, rearranged the mosquito netting, and gone off with the tin hat. Rowley lifted the netting, automatically checked Jim's pulse. He had just arrived at Little Nike after six weeks 'holiday' at Retpu, the base camp hospital.

'How is it Jim?'

'Not so bad. I'll make out, sir. The Nips playing up out there?'

'They've taken all the walking sick on a work party.'

'They've been blitzing the sick for weeks.'

'Bad?'

'Bloody awful. Major Chalmers has done his best but they've really turned on the heat.'

'They should ease up now,' Rowley said. 'The line is finished.' His professional optimism as doctor and soldier was always up front. Until they find another chore, he added mentally, keeping the cynical reality of his thoughts to himself.

Armstrong smiled his quiet smile. His big hands were relaxed on the blanket.

'The war can't last much longer, can it?' It wasn't really a question, merely a thought spoken aloud and Rowley knew he was not expected to answer. He joined Chalmers at Chauncey's cot.

Chauncey closed his eyes, then opened them as a confused muttering from one of the malaria cases disturbed him. He turned his head towards the bay where the restless man was beginning to toss and moan, then his eyes looked up anxiously at the two doctors.

'He'll be right,' Rowley reassured. 'So will you, Chauncey. Just battle on.'

'Too right, doc.' Chauncey jerked his chin slightly towards the restless man and closed his eyes as the doctors left him. By the time they reached the malaria case the man was sitting up and clawing at the mosquito net. When Chalmers caught his shoulders to ease him down he began to struggle. He tucked his head into his shoulders and tried with convulsive jerks to throw himself away from the cot.

'Easy, take it easy. You're okay here,' Chalmers said.

The man twisted his head sideways and glared up at the atap thatch of the roof.

'I'm not okay,' he squeaked. 'Get that motor bike down off the ceiling. It's going to fall. Bloody thing'll kill me.'

Rowley saw that Pinky Rhodes had headed for them from the far end of the hut, ready to add his weight to the struggle.

'Rhodes, get that bike out of the ceiling at once!' Rowley ordered. 'This man could be hurt.'

'Yes sir, at once. I'll get the ladder,' Pinky answered loudly.

The delirious man, still struggling, glared past Pinky.

'Don't need a ladder. Use your wings,' he panted.

'Of course, mate,' Pinky agreed. 'But the bike is heavy. Can't fly with it.'

'Oh.' The sick man fell back on the cot and allowed Chalmers to pull the blanket up and straighten his right arm out on it, palm up. Rowley was already on his way to the metal medical box. He returned with syringe, tourniquet, and two vials—one of quinine and one of a sulphonamide, M & B 693. As Rowley slid the needle into the vein inside the elbow Chalmers briefed him.

'He showed up on sick parade four days ago, the usual fever and rigors. Mentioned that he'd felt weak in the legs for a few days but didn't think it was enough to report sick when other men were in worse condition. Obviously not the sort of bludger who keeps trying to toss the doc. He was a little confused yesterday morning but settled down.'

Rowley nodded. 'It's going to be a long day and night. The malaria has attacked the brain.'

The malaria victim opened his eyes and let out a long howl. Then the clamour in the hospital was matched by the sound of commotion outside. The work party had returned. The doctors sprinted down the length of the hut with more haste than dignity but halted at the doorway and emerged onto the parade area with military calm.

'What the hell!'

The men were hauling two teak trunks into camp, each about two feet in diameter and some ten feet in length. Ropes had been slung around them and the men, four to a rope, strained forward like mules. Or, thought Rowley, slaves of the Egyptians hauling logs to make rollers for the massive blocks of stone that were raised into the pyramids. More than once the analogy had entered Rowley's mind as he had watched men swarming like ants on an embankment with spades and mattocks, transporting earth by the primitive pole-and-bag method, and raising timbers on to a line of bent shoulders. The tombs of the Pharaohs and the Burma-Thailand railway were built in the same way, with the blood and sweat, and lives of men.

One of the men-mules slipped and sprawled face down on the mud. The Jeep yelled and the Storm Trooper dealt the man a blow across the buttocks with a length of bamboo.

'Up, up, *speedo*.'

Rowley sprinted across the intervening space from the RAP, losing a clog. He hunkered down in the mud, neatly placing himself between the fallen man and the Storm Trooper. He expected a clout from the rear but the Storm Trooper drew back, waiting for the doctor to raise his man-mule for him. Fortunately the Jeep was superintending the progress of the other log as it slid through the mud like a sled. The prisoner, a stongly-built man no older than twenty-one, tried to raise himself with Rowley's help.

'*Speedo, speedo*,' growled the Storm Trooper.

The lad rested on one knee, leaning against the RMO.

'Legs are so weak, doc,' he said. 'They won't work right.'

The Storm Trooper grunted impatiently and the Jeep, realising that the work party around the second log was apparently idling, shouted a question. The Storm Trooper laid his stick across the kneeling man's shoulder.

'*Byoki* [sick]. This man *no shigoto* [no work],' Rowley said flatly.

Out of the corner of his eye he saw the Jeep heading back towards them. The Storm Trooper had seen too and was rumbling, ready to use his thick bamboo cane more forcefully.

'Get up. Try!' Rowley snapped at the lad, at the same time getting a shoulder under his armpit and heaving. They both staggered to their feet. Immediately Rowley moved his supporting shoulder and stepped aside. The lad staggered, his legs buckled at the knees and he was flat in the mud again. Rowley stared from the Storm Trooper to the Jeep, and shrugged.

'Hospital, this one. *Takusan byoki* [very sick]. Maybe *chorera* [cholera].'

Both the guards drew back and there was an exchange of Japanese, shrill with consternation. Before they could question the diagnosis John Chalmers arrived with Arthur Baker, Rhodes, and a bamboo-and-sacking stretcher. Swiftly they bundled the fallen man onto the stretcher and were heading back for the hospital. The Jeep scowled over his mental efforts to find a face-saving ploy.

He pointed to the hospital. 'One more man for work.'

'No men for work.' Rowley's gesture encompassed the men who were grouped around the two tree trunks, ropes slack in their hands, glad to rest. 'All *byoki*.' The Jeep took a step in the direction of the hospital but a wild and prolonged shriek halted him. The cerebral malaria case was starting to rave again.

'Mad man. That one sees many spirits . . . demons . . . *ninjas*,' Rowley said as impressively as he could manage.

'*Asoka*?' The guard's face and voice were expressionless, masking his very real fear of madness and demons.

To save face once more he would have to take the initiative, give an order. He shook his stick at the RMO. 'No men leave hospital. Maybe *chorera*. Tomorrow I see.'

It was enough for Rowley to start back to the RAP quickly before the Jeep or the Storm Trooper could take back any of the advantage he had gained. Chalmers and the orderlies already had the lad off the stretcher and on a cot. Leaving the orderlies to attend to the patient the doctors moved to the store area for a low-voiced conversation.

We could be at the Mater Hospital, Rowley thought wryly, remembering the year of residency in Sydney before he volunteered for full-time service—wearing stethoscopes as badges of rank, white coats as vestments of the medical discipline, unhurriedly intoning a ritual . . . 'I'd like your opinion, doctor . . . Thank you doctor . . .'

'He reported sick about a week ago. Diarrhoea,' Chalmers said. 'I put him on Light Duties and he seemed to be doing all right. Before you got back yesterday he told me he felt about good enough to go out again. Frankly, Rowley, I was puzzled that you put him on No Duties this morning with the Nips so insistent on having more men. What did you tell them out there?'

'Suspected cholera.'

'But . . . the symptoms aren't . . .?'

'No . . . but they're scared witless of cholera. The thought of it—and our screaming dervish over there—kept them out of the RAP. It's beriberi.'

'No oedema.'

'Dry beriberi. The symptoms have been recurring for months. Didn't Jim fill you in?'

Chalmers sighed. 'Sergeant Armstrong has been a marvel but he couldn't stay on his feet any longer. The sick parades have all been hell these last few days with the Nips demanding more men for more hours . . . putting the pressure on Colonel Anderson. He'll protect any man who is genuinely sick but he can't let the malingerers get away with it or everyone suffers. Hell, Rowley, you know there are a lot of malingerers. The tricks they think up to toss the doc.'

'I know,' Rowley said. 'But even the would-be malingerers are sick. It's a defence. And that lad is no malingerer. He has dry beriberi badly. If it affects the heart he's dead.'

Chalmers frowned. 'There are none of the usual symptoms. It's hard to make a diagnosis.'

'Yep. It's tough when men have to be belted unconscious, rotting with ulcers, or passing blood before we're allowed to hospitalise them. A man isn't sick unless he is seen to be sick by the most medically ignorant. Sick men can be branded as malingerers unfairly,' Rowley said. His mouth was compressed into a thin line.

He turned away from Chalmers and bent over the metal box. 'I'd like to go over our supplies with you, John. We're so damnably short of everything. Nagatomo makes a flying visit with one or other of the Jap doctors and we're promised drugs. But nothing arrives.'

As the doctors worked over the list of supplies they were aware of the dull thud of axes biting into wood, the long grinding note of crosscut saws pulling the length of the long, flexible blade, the gong sound that rang when the blade stuck in green wood and flexed. It rang into Rowley's thoughts, bringing sensations of a heat that was sharp and clean, not the steaming heat of the jungle: the acrid scent of eucalyptus mixed with honey sweetness from a wild bees' nest: the ring of a crosscut saw sticking in wood as men cut fence posts on the butcher's property outside Kyogle in northern New South Wales. He concentrated on his lists, determinedly suppressing thoughts that brought a wave of homesickness—a disease that, untreated, set up a gangrene of the spirit that could be fatal. The noise outside ceased.

The sound of an engine coming up the line took both the doctors outside again. The engine had come from the 154-kilo peg where the two sections of the line were to join, linking Bangkok on the east coast of Thailand with Moulmein on the west coast of Burma. It pulled one flat-top truck and three prisoners, probably the mess detail to collect the noon ration, crouched on the flat-top. Already the cooks were carrying out of the kitchen hut the buckets of boiled rice that would be a midday meal for the work gang.

The Storm Trooper and the Jeep were circling about their work party of sick men like a couple of ill-tempered cattledogs, growling, shoving, cuffing as the men strained to manhandle the two great lengths of teak over to the flat-top. But the trunks that had been dragged out of the forest had undergone a transformation; they had been squared off, tapering slightly towards one end. Rowley and John Chalmers strolled over to the track to satisfy their curiosity. Major Don Kerr, second in command to Lieutenant-Colonel Anderson, had just arrived with an ox cart full of sacks of sweet potatoes and the unpalatable root vegetable that the men called 'Hairy Marys', and Captain Bill Drower, the long, gangling interpreter, joined them. Closer inspection showed that the squared poles had each been painted down their sides with Japanese characters, ideographs which Rowley could not read although he was fairly adept in reading the phonetic characters of the Jap language propaganda sheet.

'Well, damn it all if the little blighters haven't been making totem poles,' Bill Drower drawled in his gentle Oxford accent. 'They are commemorating the completion on this day of the railway built by the courageous soldiers of the glorious Imperial Japanese Army.'

'Very good for big Nippon celebration, *desu ka*?' The Storm Trooper, who had been known to smile but seldom, was grinning broadly.

The little knot of officers stared at him blankly and the grin faded. Obviously some expressions of congratulation were in order.

'*Asoka*?' Bill Drower said. The non-committal 'Is that so?' concealed a variety of thoughts, but it satisfied ritual courtesy. The Storm Trooper could take it as complimentary if he liked.

The logs were swung on to the flat-top. The mess detail clambered aboard with their tin drums. The ration would be enough to supply each man at the work camp with a dixie of watery boiled rice. The engine trundled away along the track into the darkness of the jungle and the rain turned on again. Dismissing the work party the Nips scampered for their guard house and the sick men slopped away with mutters of relief to their billets. The camp was quiet for at least two hours, then the people in the RAP hut heard a renewed commotion as the men were routed out again. The Storm Trooper appeared at the doorway of the RAP with Bill Drower.

'Gentlemen, we are politely requested by the excellent representatives of the Japanese Emperor to participate in forming a guard of honour for certain exalted persons who are soon to pass by our humble and unworthy camp,' Drower drawled.

The exalted persons, presently greeted by a line of saluting third-class Korean privates and an assemblage of ragged, uninterested prisoners of war, were Japanese engineer officers. A small diesel engine pulling several open steel trucks of the sort that would be used for coal haulage in Australia, appeared out of the jungle from the direction of the Burmese section of the railway. For the comfort of the very senior officers they had been furnished with wooden forms to sit on and little canopies of atap perched precariously on thin bamboo poles.

No Egyptian king could have believed that he was travelling with greater pomp and ceremony, Rowley Richards thought. Such are their delusions of grandeur. And we are the slaves. When the numbers are finally counted what will be the cost of their damned railway in human lives? One for every sleeper?

That night Sergeant Jim Lynch described the ceremony at the 154-kilo peg to a group gathered around Jim Armstrong's cot. Rowley had settled down for a chat with grey-headed 'Old Silver' Armstrong, who seemed a little stronger and more cheerful. Pinky Rhodes and Arthur Baker were sitting on the bamboo platform with their chins on their knees, ribbing Old Silver about being a spine-basher.

Jim Lynch had bathed and shaved and changed his mud caked G-string for clean shorts and a shirt. The shorts were a faded pair of army trousers that had been hacked off at thigh length when the legs became too ragged to be of use. The shirt no longer possessed sleeves or buttons and had been faded from its original khaki to an insipid beige by Lynch's laundry methods: boiling up his rags in a cut-down petrol tin perched on an open fire.

The camp ablution facility was a dam built across a watercourse in a gully, and shaving razors were manufactured from the bits of metal that could be thieved from Jap trucks by their prisoner drivers. Yet Lynch carried out his bath and change ritual as meticulously as if he had come in from an afternoon's tennis at some home in Darling Point and was intending to dine with the ladies in a candle-lit dining room overlooking Sydney Harbour. An open-faced man of thirty, he was one of life's optimists without being determinedly hearty. He worked without complaining and had a keen appreciation of beauty. If a flower was blushing unseen in a desert or a gem wasting its fair rays serene in an ocean cave, Jim Lynch would be the man to discover and admire it. He had been a lawyer in civilian life.

Like other sergeants in the work kumis, Lynch was a '*kumicho*', the Japanese equivalent of a ganger or overseer. He had been *kumicho* of the party that had gone out in the pre-dawn darkness to the 154-kilo peg. He had detailed the fittest men to build the torches, actually no more than large bonfires, by which they would work until daylight. Then, cheerfully cooperating with the guards, he started others straightening spikes along the laid track so that the scene was one of industrious activity by the time the engineer officer, whom Lynch had privately nick-named 'Ananias', arrived.

According to Lynch, Ananias was as hypocritical and treacherous as the high priest of the Sanhedrin who had led the outcry for the crucifixion of Christ; an expert at driving the work force for hours and at boasting of the superior strength of his race. Lynch's legal mind was pleasantly occupied in putting the case against Ananias before a hypothetical war crimes tribunal whilst Lynch's undernourished body wrestled with a spike-driving hammer. The section of his neatly compartmented mind that remained aware of the welfare of his kumi had noted the dark shapes of men that moved stealthily over the almost intangible division between firelight and the pitch blackness of the jungle. Safe in the blackness, the sick who had been forced out to work would find a bamboo thicket somewhere, or a hollow under a fallen teak giant, to lie down. Others would weave from darkness to fire-glare, gathering fuel to keep the torches blazing while the fit spelled each other, straightening spikes or levelling the fill along the track. It was a careful saraband of survival performed against flickering orange lights. A scene from Dante's *Divine Comedy*, Jim Lynch thought. And we are the 'perduta gente'—the lost—only we haven't abandoned all hope. Not by a long shot.

Daylight came with a sudden shower followed by stifling heat. Steam rose from the earth of the embankment and the rank jungle vegetation. Ananias was fussing over the tidying up of the embankment along the 130 yards that remained between the last section of line from Burma and that from Thailand. He was giving orders for lengths of rail to be stacked on the embankment. Shortly before midday the engine pulling one flat-top that had taken the mess detail back to the 131-kilo camp arrived at the work site. Two great teak posts painted with Japanese characters lay on the flat-top with the ration buckets.

At an order from Lynch the kumi downed tools and the men began to line up with their dixies for the midday rice, but angry shouts from Ananias were transmitted to the guards, who converged on the food queue and halted the doling out of the ration. Men, Lynch gathered, were required to unload the two posts. Others were to dig holes on either side of the line at the 154-kilo peg. Apparently the posts were to be raised as some sort of commemorative obelisks. Lynch speculated on which the jungle would reclaim first; the obelisks or the narrow-gauge line that looked like a glorified rollercoaster track.

The obelisks, or totem poles, were finally erected and the earth at the bases stamped hard by naked, calloused feet. Ananias appeared satisfied. The prisoners sat or lay in groups, eating rice with fingers or with metal spoons they had made themselves. They speculated about the unfinished gap between the two sections of the railway.

'It's an expansion joint,' young Corporal Rooney volunteered. 'When the rails get real hot in the Dry, they're going to stretch out like licorice straps and the two lots will join up. Isn't that right, Stanley?' He applied to his mate for affirmation.

'Ah, Mug, you're off your head,' Bombardier Stanley said glumly. 'This rice is gawdawful. D'yer want mine?'

'Nope. Not so hungry today.'

Lynch looked across at them and frowned. Rooney and Stanley were just youngsters, healthy lads who worked as a team, scrounging for useful bits and pieces around the camps, bartering their carefully hoarded treasures with the natives for food, pooling their resources. They kept up a steady banter and had never been known to refuse food. If they couldn't tease each other or eat their tucker there had to be something wrong. Lynch tucked a memo into a corner of his mind to have them both on the night sick parade.

The purpose behind the unfinished length of line became clear when a gang of Japanese soldiers came trooping down from the Jap camp, followed by Ananias and his fellow engineers. The Jap privates first-class shouted orders at the Korean guards and punctuated them with slaps to the face and head. The guards transmitted the orders, with cuffs and kicks, to the prisoners, until the men were milling around like ants that had been stirred out of the midday somnolence of their ant heap. The prisoners were to provide the Japanese soldiers with their tools, crowbars, hammers, spikes, shovels—anything that would make a soldier look like a heroic railway builder. The Japs were going to lay the last 130 yards of the railway.

The prisoners were herded into a cutting out of sight of the Japs congregated at the ceremonial peg. Thankful for the rest and incurious about the activities of their captors, most lay down and fell asleep. Jim Lynch, feeling the gut-quaking onset of diarrhoea that occurred now every time he put food in his stomach, made a trip into the jungle. Then he made a nest for himself in the undergrowth crowning the bank above the cutting, and dozed. He felt feverish. To sleep would be wonderful but he wanted to see what the Japs were going to do.

The sun was not much west of the zenith when a train arrived at the Burma end of the line. Japanese officers descended from atap-canopied open trucks and were greeted by Ananias and his fellow engineers. A film crew began unloading themselves and their cameras from the truck at the rear. All the Jap officers seemed to have hand cameras.

Take: six Jap soldiers, carrying a length of rail, grinning widely.

Take: one Jap soldier, legs wide, straddling a spike, thirteen-pound hammer gripped resolutely, grinning widely.

Take: group of Jap soldiers watching the last rail being laid, Ananias in the foreground, all grinning broadly.

The citizens of Japan, and as much of the world as the Japanese propaganda machine and radio broadcaster Tokyo Rose could reach, would learn that the soldiers of Nippon had achieved the incredible, a railway through the Burmese jungle. But the world would remain ignorant of the terrible story of the POW slaves who built the line. They suffered and died, unable to communicate their fate.

Thank God the line is finished, Lynch thought. Things may be better now. I'd like to get out of this heat. The fever is ripping it out of me.

The rest of the afternoon was a holiday for the guards and the prisoners. The Korean guards relaxed once the high officials had been trundled off in their canopied trucks and Ananias and his engineer colleagues had departed to the Japanese camp where, from the sounds of merriment, they immediately broached the sake.

Tools were gathered and loaded onto the flat-top with as many of the work party as the truck would carry. The rest of the kumi trudged along at the side of the line. When they reached a trestle bridge over a small river midway between the 154-kilo peg and the camp, the guards ordered the train to be halted. Shouting to the prisoners that it was time for *resto* and swim the Koreans clambered down the bank and stripped to their G-strings.

Within minutes all men were in the water. Even the least fit seemed to find new energy. They splashed each other and shouted like kids out of school. The Koreans brought down a few sticks of dynamite, usually well-guarded, and after clearing all men out of the water they blew the river for fish. Guards and prisoners formed a line across the river to gather the stunned fish that floated with white bellies up.

The catch was so plentiful that there were fish for the kitchen and RAP and even rice and fish cakes in the 131-kilo camp that night. As Jim Lynch walked to the dam for his bath he saw Rooney and Stanley crouched at their fire frying rice and fishcakes. They had seemed in high spirits again at the river, he remembered, so the ill-temper and loss of appetite at midday had been a passing thing. He need not suggest that they report sick.

Lynch felt relaxed, a pleasant feeling because he had been unusually tense and ill-tempered himself for a couple of weeks. He decided to eat with Don Kerr, John Shaw and Rowley Richards. But he discovered that Rowley had eaten and gone back to the RAP to spend some time with Jim Armstrong. There was a strong bond between the medical officer and Old Silver. It had existed since the regiment, newly formed, moved into training camp at

Holdsworthy. Lynch collected some native cigars, bought by Don Kerr at the village of Little Nike, before he headed for the RAP. Rowley was not a cigar smoker. He still used his pipe but it was only brought out on special occasions now as his store of tobacco was almost exhausted and he disapproved of the strong native tobacco which aggravated the irritation in upper respiratory tract infections (URTI). After 'pay days' and the purchase of more cigars from the native traders there was always an increase in URTI on sick parade. Also, with him it was a point of honour not to bring his pipe out in front of men who had no smokes.

Having described the big celebration for Japanese to the group in the RAP, and heard their story of the events in camp, Lynch produced his cigars. Jim Armstrong did not smoke but insisted the tobacco smell wouldn't upset him if Pinky, Arthur Baker and Lynch wanted a cigar. Rowley decided he could permit himself one pipe. He was tapping shreds of tobacco into the bowl when a soldier pulled aside the mosquito netting at the RAP doorway.

'Doc, Corporal Rooney's just keeled over. He's dead. He was walking to the latrines and he just fell on his face. His mate Stanley is sitting beside him. He hasn't said nothing. Just stares at the body.'

Rowley put his pipe away. 'Tell Major Kerr that we'll need a burial detail,' he said. 'I'll come right away.' He turned to Lynch. 'Any trouble with Rooney today?'

'No appetite and very gloomy but he recovered his spirits when we were swimming.'

'Stanley?'

'He was off his tucker, too.'

'I'd say cerebral malaria. We'll get Rooney underground and Stanley into the RAP at once. They belong to your mob, Jim. Will you take the Catholic service or would you like me to do it? I learnt the drill from a priest in Australia.'

'I'll do it. Father Pete's back down the line.'

They did not get Stanley into the RAP at once. He insisted on being one of the stretcher-bearers for his mate, but he collapsed in the cemetery. Three days later Jim Lynch said the right words for Stanley as he had done for Rooney.

Rowley marked Stanley's death on the nominal roll before he put on his extra clothes to go to bed. Two apparently healthy young men dead in three days and he, as a doctor, had been able to do nothing for them. They had died in a war in which no shots were fired. The enemy was the jungle and the diseases it spawned, and the Japs were in league with that enemy in so far as they withheld nourishing food and vital drugs. He had foreseen some of the problems of warfare in the tropics right back at the beginning, but he could never have dreamed anything like the nightmare reality.

Tipping the shreds of tobacco from the unsmoked pipe back into the airtight tin that contained his tiny store, Rowley wrapped himself in his blanket and lay down on his cot of rice sacks. The line was finished. For the Japanese it was a triumph, a great engineering feat. How did he feel about it?

He had a sense of anti-climax, a draining of energy that left his limbs weak and his mind numb. Tomorrow he would start to function again, consider the troubles ahead, for there were sure to be troubles ahead, on and on, until the war was over. Of the men in his care many would survive and at the end of it all he might find out what '*s*' is—the survival factor. But that was for tomorrow. Tonight he was empty of spirit.

He had felt this way once before, aeons ago—no, one year and eight months and two days ago, on 15 February 1942—sitting alone at the wheel of an army truck, parked on the Tanglin Golf Course, Singapore. The Supreme Allied Command had surrendered to the Japanese. Singapore, impregnable fortress of the British Empire in the Pacific, had fallen. General Gordon Bennett, the General Officer Commanding, Australian troops in Malaya, had announced the capitulation at AIF Headquarters.

The acrid smell of gunsmoke was heavy on the humid air of devastated Singapore and heavy, in imagination, in every living room throughout Australia into which radio brought stunned civilians news of the unthinkable. Rowley remembered that he had felt empty, drained. There had been the same sense of anti-climax. Thank God that's over, he had thought. It's out of our hands.

All that seemed a whole lifetime away. The pre-war militia unit, Victoria Barracks, Sydney University and the declaration of war were from another universe.

2

Boys Being Soldiers

At 8.15 p.m., Sunday, 3 September 1939, the BBC Overseas Service relayed to the countries of the British Empire the voice of the British Prime Minister, Mr Neville Chamberlain, announcing the decision to make war on Germany. The British Ambassador in Berlin had informed Hitler that morning that a state of war existed. Chamberlain, obviously emotional, said:

> This morning the British Ambassador to Berlin handed to the German Government a final Note stating that, unless we heard from them by 11 o'clock that they would withdraw their troops from Poland, a state of war would exist between us. I have to tell you that no such undertaking has been received, and that consequently this country is at war with Germany.
>
> You can imagine what a bitter blow it is to me, with all my long struggles to keep peace, and yet I cannot believe that there is anything more I could have done or anything different that would have been more successful.
>
> Up to the very last it would have been quite possible to have arranged a peaceful and honourable settlement between Germany and Poland. But Hitler would not have it. He had definitely made up his mind to attack Poland whatever happened . . .

Wherever in the British Empire the radio carried his voice, the people knew that they, too, were at war. Ties of blood as well as those of trade were strong and they would not stand aside when Great Britain went to war. Hitler's invasion of defenceless Poland was not just a distant war in Northern Europe, far away and of no concern to Canada, Australasia and South Africa. It was the consequence of the egomaniacal ambition of an Austrian paperhanger, a politically astute opportunist who had connived and ranted and lied his way to the Chancellorship of Germany and at last to dictatorship, an ambitious demagogue who ignored the sanctity of national borders and lusted to rule all Europe and dictate to the world. Such men had to be stopped.

Following Neville Chamberlain's announcement the Prime Minister of Australia, Robert Menzies, made a nationwide broadcast from Melbourne at 8.45 p.m.: 'It is my melancholy duty to inform you officially that, in consequence of the persistence by Germany in her invasion of Poland, Great Britain has declared war on her . . . Australia is now also at war . . .'

It had to be. Aggressors had to be stopped somewhere. If Hitler was not stopped in Europe, if England was conquered, what then of the countries of the Empire? None was yet strong enough to survive the premature severing of the umbilical cord tying it to the Mother Country. Should England fall the Royal family might flee and the King might make his capital in Ottawa or Canberra, but the scattered brood of the Empire could not go on fighting without strong alliances.

The only possible ally was the United States of America. But who could predict what that country would do? Would they preserve neutrality until their own interests or even coasts were threatened?

As 'No man is an island entire of itself; every man is a piece of the Continent, a part of the Main', so the fall of Great Britain would diminish all those countries that were part of her dominion. On 3 September 1939, the death bell tolled for them all.

The sophisticated exigencies of international politics and economics were scarcely conceived or understood by the average subjects of the English King in all those countries that were coloured pink on the world map. But they knew they had to stand together. Wherever someone owned a wireless set, perhaps at only one or two houses in a suburban street, one or two houses in a country town, the homestead of an Australian sheep station, a Canadian wheat farm, a Rhodesian tobacco plantation, the householder and the neighbours who had no wireless were gathered to listen to the relay from the BBC.

In Australia, adults had been grave-faced all day, had neglected routine work and talked of things incomprehensible to their children, who were permitted to stay up past their usual bedtime for the evening broadcast. They heard the voice of Mr Chamberlain, thin, old, and weary, barely audible through the crackling static.

Afterwards fathers poured beer or sherry for the neighbours if the family could afford the luxury. The Depression still affected the average income-earner. Mothers went to make pots of tea. Fathers talked in very low voices, as if someone they knew had died, and mothers cried or looked as if they wanted to cry. Children wondered why. They were excited because war was a cinema adventure on Saturday afternoon, with knight crusaders, Ivanhoe, Beau Geste in the Foreign Legion, aeroplanes dogfighting in 'flicks' about the Great War. It was an exciting departure from routine. Children were sent to bed and lay awake wondering if they would have to go to school tomorrow or if there would be no school until the war was over.

Adults did not go to bed. They sat around the radio talking about people they knew who had been killed in the Great War, and waiting for the broadcast by His Majesty the King, which was to be made at 3 a.m., Sydney time. Enunciating carefully, so that his slight stammer was perfectly controlled, the King put into words the truth that civilised, responsible people know in their hearts: that there comes a moment of decision when action must be taken against bullies and aggressors.

In this grave hour, perhaps the most fateful in our history, I send to every household of my people, both at home and overseas, this message, spoken with the same depth of feeling for each one of you as if I were able to cross your thresholds and speak to you myself.

For the second time in the lives of most of us we are at war. Over and over again we have tried to find a peaceful way out of the differences between ourselves and those who are now our enemies. But it has been in vain. We have been forced into conflict for we are called upon without allies to meet the challenge of a principle which if it were to prevail, would be fatal to any civilised order in the world. It is the principle which permits a State to disregard its treaties and its solemn pledges, which sanctions the use of force, or threat of force, against the sovereignty and independence of other States . . .

At Dapto, near Wollongong, on the South Coast of New South Wales, a militia unit, the Artillery Survey Company, from Eastern Command Headquarters at Victoria Barracks, Sydney, had just moved into training camp for summer manoeuvres. The rows of tents had been pitched, HQ, camp kitchens, communications centre, ablutions area and latrines established. The camp was on the greyhound racetrack on the outskirts of Dapto shopping centre, which amounted to a single street with post office, public hall, pub, a few shops and a picture show.

A dance was held in the public hall on Saturday nights and when permanent army or militia men were in camp it was the place for soldiers to try to meet the bold-timid girls who wore cotton frocks with demurely rounded necklines and puffed sleeves, and ribbons in their hair. The girls' mothers had warned them 'about soldiers' and the bold young men in uniform, who were really mostly quaking kids inside, trying to be soldiers, were too shy to ask the girls to dance. Boys and girls gathered on opposite sides of the dance floor. Girls danced with their regular boyfriends or with each other. They drank lemonade or orangeade.

A girl who came to the dance unescorted and danced with a strange soldier was considered a bit fast and the other girls whispered about her. A boy in uniform who plucked up enough courage to cross the floor and ask one of the local girls to dance was entitled to boast, and was admired and envied by his mates. But the fastest girls knew better than to be whizzed off into the shrubbery and the boldest boys in uniform were scared too. So the young soldiers slipped outside to drink beer, which increased boasting but not courage, and drifted off in twos or threes or along to the picture show.

Some wide-awake types gave their mates the slip and were vocal next day about the good time they had had with a local sheila: actually they had walked back to camp, killing time on the way by sitting under a gum tree and communing with the stars. The Southern Cross constellation could keep the secrets of the boys who were trying to be soldiers and were frightened of the mysterious beings called girls. Their ages ranged from eighteen to twenty-two or twenty-three.

On Saturday night, 2 September 1939, the boys in uniform who had attended the local dance were not as interested as usual in the girls on the other side of the hall. Elusive adventure seemed almost within reach. Although it was only months since the British Prime Minister had talked confidently about 'peace in our time', there was now talk that war was imminent. War was the big adventure, for which they were trained and ready . . . although perhaps not one of them had ever witnessed sudden violent death.

After Saturday night came Sunday morning, which was no different from any other morning for the local dairy farmers who were working by artificial light each dawn. Most of the farms in that lush, gently undulating, semi-rural area were dairy farms. Even families with a large block of land beside the house would keep a cow. For the non-farmers Sunday morning was sleeping in, reading the *Sunday Sun* or *Truth.* In the Sunday paper comic strips Ginger Meggs racked his brains to avoid the bullying Tiger Kelly; and Bib and Bub, the gumnut babies, continually fled from Bad Banksia Men; but on Sunday morning 3 September the talk was of possible war, not Ginger Meggs.

The Artillery Survey Company moved into camp at Dapto that morning, having missed the pleasure or pitfalls of the Saturday night dance. Although they were too busy establishing camp to sit around with Sunday papers, there was continual talk, rumour and speculation about war with Germany. The very young men were excited. Men old enough to have fought in the Great War or to have been boys who became the men in their families when their fathers went away and were killed, were tense and grave. Tension increased as the night fell and darkness blotted out the spectator stands of the greyhound track. The headquarters room was the focus of attention. And by 9 p.m. it was official. They were at war. The sad faces of the older men showed that they were reliving memories of the war of Gallipoli and Flanders. The young militia men generally felt that it was all a great lark. Now they would be able to show how well they could do all they had been training to do for years.

The commander of the Sound Ranging Group of the Artillery Survey Company was a twenty-three year old First Lieutenant, Charles Rowland Bromley Richards. He was as excited as any other young man by the idea of being a soldier in a real war. The greater part of the world had been at peace since 1918, the mud of trenches and the lung-searing horror of poison gas was a muttered tale without meaning; the diggers marching proudly on Anzac Day with medals on their best blue suits were a brotherhood of heroes; heroism had outlasted horror and war was a glorified adventure in the best *Boy's Own Paper* tradition. It was an interesting theoretical problem and Rowley loved solving problems. He also loved putting theories to practical tests.

Rowley had always been independent in his thoughts and in planning his life. His family background was unusual as his parents were deaf and dumb. Charles Richards, his father, had been born deaf, the result of German measles contracted during his mother's pregnancy. Rowley's mother, Clive (a Christian name she would never explain), had caught diphtheria when she was five years old and survived the white death that claimed so many children, but the complication of an inner ear infection deprived her of hearing. Neither of these strong-minded, intelligent people allowed their infirmities to interfere with their lives.

Charles was schooled at the School for Deaf and Blind Children at Darlington, near Sydney University, where a little girl named Clive Bromley was also a student. Charles learned to lip-read, trained as a mapping draughtsman, and found employment with H.E.C. Robinson, a leading Sydney firm of cartographers. Clive Bromley became a teacher of handicapped children at the school at Darlington and was even more expert than Charles as a lip-reader. Each recognised a kindred spirit in the other, and they married on 24 April 1915, the day before the Anzac landing at Gallipoli.

The Great War was still in its first year. No one believed it could last more than a few more months. Charles could not become a soldier but he was hectically busy as a skilled cartographer. Clive continued her teaching and they acquired a house at Summer Hill, a suburb of quiet streets and spacious gardens. Many of the houses, the Richards' included, had tennis courts so the young couple played tennis on summer evenings and Saturday afternoons with other young marrieds. Although many of the husbands were already in uniform, war was far away until tennis friends disappeared from the Saturday parties and were later listed among the killed or missing in action. Clive knitted khaki balaclavas and socks for the troops, then began alternating her knitting for the war effort with the knitting of baby clothes. When the Richards' first child was born and Charles was told that he had a son he said, 'And he shall be a doctor'.

Young Rowley went through the stages of wanting to be a tram or train driver, because you got free rides. But by the time he was ten he had decided quite independently of the elder Richards' prophecies that he did want to be a doctor. Disease had deprived his mother and father of their hearing. His younger brother, Frank, suffered from bronchitis. So the study of medicine, particularly with the object of preventing sickness rather than attempting to minimise the damage, seemed to him the greatest thing a person could undertake. He was encouraged, not only by his parents, but by a new and greatly admired uncle. Aunt Linda, his father's young sister, married Arthur Lewis, a doctor who practised in the country town of Kyogle on the far North Coast. Kyogle was to become a second home for both the Richards boys.

Life was good. Rowley and Frank played tennis after school, shared their father's enthusiasm for cartography and mathematics, and Rowley spent the long summer holidays with Aunt Linda and Uncle Arthur.

For an eleven year old boy, travelling alone to Kyogle was an adventure. The steam train left Central Station, Sydney, in the early evening. Suburban backyards fled by to the rhythmic clicking of the wheels, drifts of engine smoke curled past and the box compartments of the carriages were fascinating little rooms with plush covered seats, shining brass luggage racks, brackets between the windows to hold metal-stoppered water-bottles, thick railway glasses, and framed pictures of rural scenes. The suburbs gave place to tall timber and little, brown and green painted country stations where the locomotive sighed clouds of steam. Tea from the railway refreshment rooms was hot, strong and served in thick white cups. Ham sandwiches on dry white bread could be bought with the tea for sixpence. At Casino, Rowley had to change from the steam train to a single-carriage rail motor which hummed and lurched the last twenty miles to Kyogle. He was helped with his luggage by Norm Larkin, a recent veterinary graduate from Sydney University who was going home to Kyogle for Christmas. Within the first couple of miles out of Sydney he had learned that the cheerful, young traveller whom he christened 'Smiler' was the nephew of Dr and Mrs Lewis, who lived across the road from his father.

He briskly took charge of Rowley, made him as comfortable as one could be for such a long rail trip, and plied him with information on the country town and its inhabitants. He was still yarning when Rowley, already feeling like a native of Kyogle, dropped off to sleep. Norm wakened him for the change to the rail motor.

Kyogle fulfilled all Rowley's expectations. The wide main street, dressed with coarse river gravel, and the weatherboard buildings with pillared verandahs shading the footpath, were like a western movie set. Uncle Arthur introduced him to the local butcher, who kept a couple of horses for his older boys and offered Rowley the use of the spare horse for riding. It was a bit rough as a riding hack but Rowley loved it. The butcher also owned an abattoir out of town and suggested that the boy might like to go out and help draft the cattle into the yards. Soon he had made friends with the police sergeant's daughter, and the bank manager's son, Des Makepeace, a few months older than himself. The three went riding almost daily and swept back down the main street of Kyogle at a brisk canter at dusk, with Rowley, small and light, bouncing on the butcher's horse between his two companions, convinced that he was tearing along at a gallop. He supposed they shouldn't be galloping in the main street, with gravel spurting out from flying hooves and as if they were about to shoot up the town, but it had to be all right as the police sergeant's daughter was the ringleader. Kyogle provided more excitement on Christmas Eve. A grass fire started somewhere behind the main street, took hold quickly in long summer grass and tinder-dry weatherboard buildings and half the town was burnt down. A few years later, when he was a senior student at Fort Street High, and spending his summer holidays at Kyogle as usual, the other half of the town burnt down.

By that time the economic tremors that had started with the Wall Street crash in 1929 had built up into the full-scale quake of the Great Depression. Charles Richards was never out of work as old H.E.C. Robinson succeeded in keeping on all his draughtsmen for two or three day shifts a week, but money was very tight. The Richards rented out their tennis court and Rowley and Frank prepared it after school. Des Makepeace helped them. His father had died from complications of appendicitis and he and his mother now rented part of Rowley's grandmother's home, which backed on to the tennis court.

Tennis, and riding when they were at Kyogle, were Rowley and Des' sports, and since there was no money for the pictures or dances Saturday night was spent doing the crazy maths problems in the *Saturday Evening Post.*

At the end of primary school, Rowley had organised himself into applying for enrolment at Fort Street Boys' High School, a selective school, and was sure he had been accepted only because he lived within the area. By Fort Street standards, he thought, he would be at the tail end of acceptable scholars. If he didn't qualify for medicine at university he would become an actuary, but medicine was his love.

In the Depression years the other obstacle in the way of his attending university was the question of fees. Always planning ahead, Rowley devised various schemes to work his way through Uni. But Uncle Arthur Lewis, convinced that his nephew by marriage would make a good doctor, had some money put aside; he offered to pay Rowley's university fees, the money to be repaid when Rowley set up in practice himself. Rowland Richards graduated from Sydney University, Bachelor of Medicine and Bachelor of Surgery, in December 1939.

Rowley's life during the university years had been divided into two separate strands, the study of medicine and training in the militia Artillery Survey Company at Victoria Barracks. He could have joined the Sydney University Regiment, but reasoned that he wanted to get to know the type of people he would be treating if he set up practice in a country town like Kyogle. So, with his interest in maths and his home background in cartography, it was natural that he should join the Survey Company. At university he was a student learning to be a doctor; in the militia he was a gunner learning surveying, sound ranging (fixing the position of enemy guns by sound), and flash spotting (fixing the position of enemy guns by sighting the flash through a modified theodolite). On 4 September 1939 he was gazetted as a one-pip lieutenant.

The senior officers in his company had volunteered to join the AIF as the war rumours strengthened, so on the first day of war Rowley, as the most junior officer, found himself under canvas near Dapto and in command of the Sound Ranging Group which was shortly to be formed into a battery. The locals, who seemed to expect an immediate attack on Australia, looked on the young soldiers as having been posted there to save them from the Huns or their Italian allies.

The army did not require Rowley Richards as a full-time soldier immediately and his civilian medical life continued for another year. He became resident medical officer at Sydney's Mater Misericordiae Hospital, having done his student studies at Sydney's St Vincent's Hospital. But Rowley's experience of hospitals went further afield than the great Sydney hospitals. When he was in his first year at university, Uncle Arthur had decided he should have practical experience in a country hospital and planned to introduce him to the wards and operating theatre of Kyogle Hospital. But before Uncle Arthur's plans could succeed, Rowley had to contend with the hospital's guardian dragon. Matron Nora McKidd, cold-eyed and warm-hearted though contemptuous of medical students, told him that she had no time to dry behind his ears and decreed he would have to learn to scrub out the theatre. The test was passed to their satisfaction and when he reached his final year Rowley had administered some fifty anaesthetics and assisted his uncle during several operations.

While Rowley was doing his residency at the Mater, the military machine had moved into gear and was picking up speed. All over Australia the lives of young men were disrupted as they left farms and agricultural colleges, university engineering courses and factories, building sites and banks, mines and legal offices, printing works and art schools to join the AIF. There was no conscription for overseas service; they were all volunteers and war was a great adventure.

But the military machine was without glamour and did not discriminate. It was built to dehumanise, to transform men into components, cogs, rods and pistons that functioned to order. Civilians were taught from childhood 'Thou shalt not kill'. Soldiers had to be taught to kill reflexively, without thinking. The recruits, so many of them boys of eighteen, were stripped to the pelt, subjected to the indignities of physical examination, issued with ill-fitting khaki uniforms, boots, slouch hats and rifles. Infantrymen were sent out onto parade grounds with bayoneted rifles in their hands and taught to disembowel dummies of straw and rag tied to posts. 'Shove it in, soldier,' sergeants bawled. 'Harder, you're not cutting cake. Now rip, down, sideways. Give it a good twist or you won't get the bayonet out.' The enemy wasn't another human being. The enemy was a straw dummy. The enemy was a collective monster that raped women and killed children unless it was killed first.

In artillery regiments the gunners found the range, loaded the shells, jumped back from the recoil, found the range, loaded, jumped back again, and cheered when the target went up in a cloud of debris and dirt. The target was The Target and the debris was the destruction of The Target, not tatters of bone and blood-stained flesh that had once been human like themselves. It was a coldly calculated dehumanising process but in wartime it could not be otherwise. The military machine could not function unless its components responded automatically, and soldiers who stopped to think were dead soldiers. They were being trained to survive in battle.

For the many who wanted nothing in life but to obey orders, no matter whether the bosses were civilians or officers, the discipline was good; for the individualistic few it was a trauma that would affect them for all their lives. But they had to learn to kill, to fight, to survive.

The 6 Division of the AIF left Australia in January 1940 to complete its training in Palestine. Following the signing of the Pact of Steel between Hitler and Mussolini in May 1939, war with the Italians in North Africa appeared inevitable. Fighting over an African desert might make little sense to men enlisted from the Southern Hemisphere (although volunteers were not expected to 'reason why'), but Germany was ready to invade France, Belgium, Holland, Denmark, Suez and Norway and most of Middle Europe.

Britain's First Lord of the Admiralty, Winston Churchill, belonged to the military school of thought, as old as Julius Caesar, that felt the Mediterranean was the tactical key to victory in warfare in Europe. Losing Africa would be losing the Suez and the Mediterranean. It was not for Commonwealth troops to reason why.

For twenty-two years the eleventh day of the eleventh month had been, for Australian school children, a day of morning gatherings in quadrangles or assembly halls, singing 'God Save the King', 'Rule Brittania', and 'Advance Australia Fair', speeches by heads of schools and standing silent for two minutes at eleven o'clock to remember the fallen of the Great War of 1914-18, which was fought to end all war.

Rowley had been two and a half years old on that first Armistice Day. On Armistice Day 1940 he was in camp with the Militia Eastern Command, Army Service Corps, Liverpool, on the fringe of Sydney, having transferred to the Australian Army Medical Corps to enlist in the AIF during the previous September. As a second-lieutenant in the militia he had commanded the Sound Ranging Battery of the Horse Field Artillery for twelve months and was an artillery officer in his own right. He had received a message on the previous day from the Commander of the 8 Australian Division Artillery, Brigadier Callaghan, that a new artillery regiment, the 2/15 Field Regiment, Royal Australian Artillery, had been formed and he was to be appointed its regimental medical officer.

Rowley was delighted that he was a real soldier in the AIF at last. His breeches were sharply creased, leggings and belt polished to mirror finish, slouch hat at just the correct angle for an experienced military man, when he presented himself to his commanding officer, Lieutenant-Colonel John O'Neill, at Division Headquarters, Rosebery Racecourse. O'Neill was no stranger. He had been a major in the 2/9 Field Regiment and was a friend of Lieutenant-Colonel 'Pete' Davies, Professor of Physiology at Sydney University who had introduced Rowley to him. Rowley liked and respected O'Neill. He had a round, cheery face with peaches-and-cream complexion against which his black bushy eyebrows and black moustache were a startling contrast. Short, well-covered and dapper, in civilian life he had been sales manager for Kellogg's in Australia.

The other officers at the meeting had all been militia men. Major Jack Meyers, one battery commander, was a manufacturer of neckties; Major Frank Ball, the other battery commander, was a banker; Lieutenant Theo Le Maistre Walker, the Intelligence Officer, was a lawyer. He was tall, had the lean and hungry look of men who think much, and Rowley thought he would make an ideal IO. All were Field Artillery men, and like Colonel O'Neill, expressed satisfaction at having a real gunner for RMO. At that stage Rowley was still thinking of himself as an artillery officer rather than a doctor. He even wore the diagonal red over blue colour patch of the Artillery rather than the brown Medical Corps patch. He was happy to be among colleagues with solid military training and proven organisational ability in civilian life.

Establishing a new regiment from the beginning was obviously going to take a lot of that ability. There would be running around and plenty of louse-ups before the thing functioned. To begin with, the regiment was to move into Ingleburn camp on 15 November. On that day the nucleus from Rosebery

Racecourse and the first detachments from the training depot would arrive at Ingleburn. That gave Rowley four days to move the few miles from Liverpool and prepare for their arrival.

'The RMO can start housekeeping again,' John O'Neill remarked.

It was a sly reference to Rowley's exploits at Liverpool during the previous two months while waiting for a posting in the AIF.

Liverpool camp had been one hell of a mess; built prior to 1914, occupied by the AIF during World War I and allowed to fall into disrepair thereafter. The wooden floors were, to describe them kindly, well worn. There were cracks between the floorboards so wide that straw from palliasses had fallen through into the space under the huts, been covered with the dust and dirt of years until finally there was a solid rotting mass, right up to the floorboards, that would have made good humus for a market garden. Rowley had promptly set about organising cleaning details to remove the refuse, literally tons of it. The men under his command moaned, groaned and tried to claim urgent business elsewhere until they found that coins as well as straw had fallen through the gaping boards. As the return varied from several shilling to pounds a day, and a soldier's pay was 5 shillings a day, cleaning became a lucrative business. Cleaning up the garbage that had been pitched from the camp area onto the river bank created vocal protests, but Doctor Richards was determined to make hygiene part of the way he fought his war. Armies might march on their stomachs, but stomachs were only strong enough to march on if they weren't upset by unhygienic conditions. The vigour with which he put his convictions into practice already had him at the head of the 'shark list' with some of the units at Liverpool camp.

Ingleburn was not quite so bad as Liverpool but was bad enough, Rowley decided. The room which had been allotted as his regimental aid post had been unoccupied for some time. The wooden tables were covered with grime and grit and rubbish had collected on the floor.

'Well, this is another dirty damned hole,' Rowley told himself aloud. His back was to the door.

'Sir . . . excuse me . . . but would you be the MO or the RMO, sir?' a gentle voice asked quietly.

Rowley spun around. He saw a grey-haired solidly built man of medium height who looked uncomfortable in his uniform and wore his slouch hat very square instead of at the angle affected by experienced soldiers. He looked bewildered and almost sheepish.

'The RMO. What's your business?'

'Sir . . . I'm Armstrong, Corporal. I'm your medical orderly.'

Good grief, Rowley thought, a new recruit and looks as if he'd turn and run if I said boo. Instead he said, 'Well, this is your RAP, Corporal Armstrong. To start with see if you can clean it up a bit.'

Armstrong looked as if the sight of the mess saddened him considerably. 'It isn't much good, is it, sir? . . . I . . . I'll need some cleaning things. Could you direct me to the quartermaster's store. I'm new here.'

'So am I, Rowley told him dryly but gave the necessary directions. He saw

Armstrong pull out a handkerchief and dab at what seemed to be a tear on his cheek and was aghast. Surely he's not crying already. I don't believe it! The man's not an eighteen-year-old kid!

Rather than witness his medical orderly burst into tears over the realities of army life, Rowley made a hurried exit and set out to discover in what other ways Ingleburn Camp was lacking in hygiene. Flyproofing for the kitchens was lamentable, the ablutions facilities looked as if they were lying in wait to spread tinea, and the latrines needed a thorough disinfecting. By the time Captain Richards was ready to return to his RAP he had a list of jobs to be done that would make him a 'king shark' once more and set the gun sergeants in a flutter over the way he took their precious crews away from the all-important gun drill . . . not that there were any guns with which to drill.

The state of the RAP astounded him. In only a couple of hours Armstrong had not only swept out the room; he had, with brush and sandsoap, scrubbed the floor, the tables, the legs of the tables—in fact everything that could be scrubbed—so that the wood looked almost new. He had acquired chairs and benches, and extracted from the quartermaster a blanket to cover the table. A big wicker pannier with a red cross on top was placed on one of the benches. Armstrong was certainly not afraid of work.

Rowley looked at his medical orderly more closely. The man's arms, exposed as he had rolled up his sleeves to work, were muscular and his hands were huge; ham hands that should have been clumsy but were surprisingly deft and gentle as he checked over the contents of the hamper.

'Well, this looks like home,' Rowley said. 'Congratulations. What did you do in civvie street?'

'I have a chicken farm, sir. At Point Claire . . . that's on Lake Macquarie, you know. Eggs and day-old chicks.' Armstrong set down a bottle as gently as if it it had been a day-old chick.

'You have someone to look after the farm?'

'The wife . . . she's one in a million. And three boys.'

'You're fortunate to have three sons, Armstrong.'

'They're not . . . yes, they . . . I mean . . . well, we weren't fortunate enough to have kids of our own so we took the boys to foster. Great kids. We adopted them. They really needed looking after.'

If I ever saw a man with a mothering instinct you have it, Rowley thought.

'Well Armstrong, from now on you're in charge of the housekeeping. I'll handle the organising,' he said. He saw that Armstrong was again mopping at his cheek. 'By the way, Corporal, what's wrong with your eye?'

'It waters. Not all the time but it does have bad spells. Looks like I'm crying like a kid sometimes,' Armstrong told him.

'Let's have a look.'

Rowley turned down the lower lid for examination. 'A partially blocked tear duct. I'll get you some drops that should control the watering. If they don't help a small operation performed under local anaesthetic can be done to clear the duct. We'll have to get a supply of drops in . . . and supplies of a heck of a lot of drugs. Your pannier looks light on.'

Everything was in short supply: drugs, bottles to store and dispense drugs, even the eighteen-pounder guns were in short supply. The 2/15th was an artillery regiment trying to train gunners without having any guns. The military machine was moving but was still in low gear. At one end of the production line top brass were evolving strategy, and at the other end drafts of raw recruits were falling over their boots and waiting to be trained while sergeants fumed and suffered from frustration. Guns were so rare that gun crews were trained with a broomstick laid on the ground ('See that, that's the barrel'), and a rock placed either side of the stick ('Those rocks, Gunner, are wheels').

Rowley set himself to looking after the supplies and Corporal Jim Armstrong set out to look after the RAP and Rowley, who had aroused his mothering instinct. The RMO seemed a pretty game and independent young rooster but he looked so very youthful and slight. Armstrong felt like tucking him under a wing.

Armstrong was what people generally call a simple man without realising that it is not a simple thing to reduce the complexities of life to a few uncomplicated principles which allow a man to live at peace with himself as well as with his neighbours. He was compassionate, gentle and loyal and had been fortunate enough to meet a girl with the same qualities so that his marriage was built on the rocks of love and loyalty. He was loyal to his community and his country, for which he had a deep love, and to personal convictions on human dignity, so he saw it as his duty to enlist. But he had an abhorrence of taking life although he was not a conscientious objector in the manner of some of the religious sects that were clamouring on a platform of the Sixth Commandment. Jim's natural loathing of violence was the reason for his request for assignment to medical orderly duties. During his short stay in the training depot he had become known as Old Silver, a term of endearment that endured among the troops, who soon took note that Old Silver was the Baby Doctor's staunch retainer and champion.

As well as Old Silver, Rowley acquired another retainer who was determined to take care of him. Assigned to him as batman, Sid Brien was a quiet, conscientious, not very robust man in his mid-thirties who walked with a limp due to arthritis in the hip.

Sid was extremely proud of his ability to polish leather and became affronted and upset if Rowley happened to admire the polish on the Colonel's Sam Browne. He then expended even greater effort to ensure that his officer's leather was better than the Colonel's. Rowley was so busy that realisation crept on him slowly that he was being fussed over and taken care of. His independence reared up with indignation until he discovered that Armstrong and Brien were hurt; so he pretended meek submission to fussing—and learned to enjoy it.

On Christmas Eve, when the Regiment had settled in at Ingleburn and was gradually beginning to function as a regiment, a message was sent from RAA Headquarters that it was to move a few miles to the manoeuvre paddock just outside the School of Artillery at Holdsworthy, and the move was to be effected by 1 January. Eight days!

The staff captain who came to Ingleburn to make the arrangements was

Fred Martin, another of Rowley's associates from the militia days. He called for a conference of the CO, the battery commanders and other HQ officers. The Quartermaster, RMO, Intelligence Officer, and Survey Officer were also to attend, and when everyone was assembled Captain Martin looked around anxiously as though he thought someone was missing. Finally he turned to Colonel O'Neill and commented that the RMO was late.

'Our Regimental Medical Officer is present,' O'Neill told him gravely. 'Captain Richards.'

'Rowley's a gunner officer,' Martin burst out. 'I've known him for years, and he's wearing our colour patch.' When assured that Rowley was indeed RMO, having graduated the previous year, Martin added: 'No doubt he is living up to the old Artillery adage, "Once a Gunner always a Gunner" '.

Christmas leave was out for the officers. Rowley and Des Makepeace, who was now a lieutenant and the acting Survey Officer of the 2/15th, went over to do a reconnaisance of the area. Most of it was under scrub with a few substantial small trees. There was no fixed accommodation in the form of huts. The Regiment was to be allowed one water point and adequate tentage.

'Anyhow, it's good practice at establishing a camp from scratch in emergency conditions,' Des commented. 'We'll survey the layout now, eat dinner, then see about rustling up the equipment we'll need.'

'At least we don't have to clean out the muck of prior residents,' Rowley told him, adding happily, 'this camp is going to be strong on hygiene, Des.'

They began the planning of a stormwater drainage system, effluent drains, grease traps, ablution areas with absorption and evaporation trenches, paths and tent lines.

The scrubby area of the manoeuvre paddock was cleared by dragging with a steel cable run taut between two gun tractors. The scrounging talents of the troops produced materials for building that were not available from the QM stores and their officers refrained from questioning the origin of items like galvanised iron and chicken wire. Theirs not to reason the difference between straight-out thieving by the men and involuntary contribution to the war effort by local farmers. The efficient flyproofing of the kitchen, the ingenious construction of the storeroom for meat, butter and other perishables (double walls of double thickness of brushwood laced into chicken wire with a space between the walls to allow cooling air to circulate), and the hygienically constructed ablutions area were Rowley's delight. He and Des were almost purring by New Year's Day 1941 when the Regiment moved into its new home, facilities had been completed, neat lines of tents erected, paths made and even laid with gravel.

'Gawd strewth, I hope we stay here for the rest of the war,' said the 'Dodger' (Artful Dodger as in Dickens because he combined a talent for dodging any form of work with accomplished thievery). 'I don't want to build no more camps with that Lieutenant Makepeace and the Doc buzzing about me ears.'

'Didn't see you doing much hard yakker though,' said his mate Billy (Bunter), whose barrel body had a capacity for consuming enormous quantities of beer.

'Dodging is an art, mate,' the Dodger told him. 'Yes, a real art.'

But he didn't manage to dodge the Doc's hygiene parades and was one of the most vehement protesters when he had to stand in the ablutions area in nothing but his skin, submitting to an inspection of ears, armpits, scrotum and toes for signs of skin diseases. 'Being a soldier don't give a bloke no privacy at all,' he complained bitterly.

Rowley was seconded to the School of Hygiene and Tropical Medicine at the Sydney Agricultural Showground the week after the move into the new camp. To his mild surprise he passed the course with distinction and managed to soak up extra-curricula information like a cerebrating sponge. He would sit at the feet of anyone with knowledge to impart.

Even his favourite sport, riding, was combined with the pursuit of knowledge. Riding afternoons with Professor Pete Davies were a combination of pleasure and medical conferences. Lieutenant-Colonel Pete Davies, the Professor of Physiology at Sydney University, was commander of the militia Field Ambulance then in camp at Ingleburn. Rowley knew him both in the militia and at the university and as a gunner sergeant and a third-year med student in 1936, the year he did physiology, was convinced that he passed largely due to Prof Pete's turning a blind eye to his shortcomings. The professor, in fact, never turned a blind eye where slackness or incapacity in medicine were concerned and had passed him on his merits. Rowley was a student of whom he expected much.

In January of 1941 the Field Ambulance was in the process of being converted from a horse-drawn unit to a mechanised one but Pete Davies was still carrying out manoeuvres on horseback and invited Rowley to accompany him. They also indulged themselves with some social riding and were supplied with mounts from the Remount Depot where Major Norm Larkin, Rowley's friend since the first trip to Kyogle, was the officer in charge.

Pete Davies needed a big horse. Although he was of medium height he was rotund, almost obese, being a gourmet and connoisseur of wines, particularly good port. His face was round, florid, endowed with an extra chin, and his scalp was a polished, bald dome. Permitted the honour of riding the General's charger, Sally, he and Rowley rode through the artillery range and dropped in at the mess of the 2/9 Field Regiment for a beer.

'Someone I'd like you to meet,' Pete told Rowley. He introduced the RMO of the 2/9 Field Regiment, Captain John Loewenthal, a young man about three years Rowley's senior. He was stocky, square faced, with a slightly swarthy complexion and rather brusque in his speech, not through any dislike of his fellow human beings but because his mind, already regarded as brilliant by his instructors and older colleagues, was totally occupied with the acquisition and furtherance of medical knowledge. He was considered to have a great future in medical research.

Riding sessions for Rowley with Prof Pete became a regular relaxation and exchange of information. Through mid-summer they rode in the Black Forest, south of Holdsworthy on the road to Helensburgh, which was the original Cobb and Co coach route from Liverpool to the South Coast, exploring for old settler's wells in a heat haze heavy with the oil of eucalyptus and loud with the rasping of cicadas. On the way back they dropped in for a beer with John Loewenthal, who shared Rowley's disdain for wasting time on non-essentials.

Loewenthal had established in the 2/9 Field Regiment a system of training and equipping battery medical orderlies, although such orderlies were not officially on strength. He talked briskly and succinctly about his system. Rowley like the idea and even felt that it could be extended to troop level with unofficial orderlies trained in first aid for each troop.

Although war action was concentrated in Europe and the Middle East and Britain was taking a pasting from the *Luftwaffe*, there were rumours that the war would extend to the tropics, probably Malaya, if Japan, which was still talking politely to America, decided for the Axis alliance of Germany and Italy. Training of RMOs at the School of Tropical Medicine had reinforced the rumour.

In early February the liner *Queen Mary* departed Sydney for Malaya, carrying a brigade group of all arms.

'There'll be skin problems, and malaria and dengue to deal with,' Pete Davies warned. 'It's tropical warfare for the 8th Division.'

The military machine was speeding up the training of its components to fight, building up the stress that transformed undisciplined civilians into cogs strengthened and tempered like steel. But the soldier doctors knew that men were not steel, but infection-prone flesh and viscera, with nervous systems that could be attacked by a virus or sent quaking into shell shock. These poorly defended masses of tissue were going to have to be trained to stay alive in an unfamiliar environment inimical to man, particularly to white man, who had no resistance to its arsenal of disease. But convincing a gun sergeant or battery commander that tinea detection was as important as gun drill, Rowley discovered, should have been added to the labours of Hercules.

At its inception the 2/15 Field Regiment was made up of the Regimental Headquarters, two batteries each of three troops. It was later restructured into three batteries: the 29th and 30th had two troops each (Ack, Beer, Charlie and Don); the new battery, the 65th, had three troops (Edward, Freddie and George). Each battery was commanded by a major with a battery captain as 2 I/C. Each troop of 120 men was commanded by a captain. A gun sergeant was in charge of each gun and there were four gun sergeants to a troop, although not yet four guns, which were in shorter supply than trainee gunners. Because battery officers were obsessed with the idea of training which to them meant the training of gunners exclusively, every man-hour spent other than in gun drill was a hell of a waste of time. Gun sergeants, being in charge of gun drill, were very important people. When Rowley obtained Colonel John O'Neill's permission to run a course of first-aid instruction and nursing training, the shock wave that started with the battery commanders became a fresh series of explosions by the time it reached the gun sergeants. Twenty-four gun sergeants bellowed like kids who had lost their lollies, only their command of language was much more colourful.

'A man from each gun and each bloody observation post and from the bloody battery headquarters as well, not to mention signals and those offspring of the Borgias that run the kitchens. That misbegotten miniature medico has got to be off his hygienically packed noggin. Does he think we're going to give the gun an enema before we shove a shell up its end?'

'Watch your language. He's going to train 'em to hold your nose and give you a dose of castor oil because you have a dirty mouth.'

'The cooks are going to be taught to shove anti-sex dope in our tucker when we get to the front so's we don't have naughty dreams or try to knock off the nearest fencepost.'

Shock waves travelled up again from the twenty-four gun sergeants to the six troop captains, to the two battery captains who transmitted them to the two battery commanders, and finally the Colonel was confronted by two upset majors. John O'Neill's black moustache and bushy black brows quivered against his peaches-and-cream complexion.

'I have considered Captain Richards' proposition and find it commendable. In other words, it's a damned good idea and we're going ahead with it.'

In the RAP, Rowley slapped his hands on his knees. 'That's that,' he told Armstrong briskly. 'We can get started.'

Finally, sixty-seven men were allocated to the course which lasted for six days and covered both the St John first aid course and the army medical orderlies' course. The gunners began to think it all might be a good idea when they found that Rowley had acquired some nurses from the Casualty Clearing Station to assist in training of the men in bed-making, taking temperatures and sponging down patients.

The Dodger's reaction to being chosen to volunteer for his gun crew had been 'Why do they always have to pick on me?' But he decided it wasn't so bad if you could get to pat a neat female rump whilst learning to mitre the corners of a blanket. The owners of the rumps had other ideas and the Dodger got several kicks in the shins. When he found that the St John course involved a bit of real hard study he developed a history of migraine headaches which kept him away from training sessions. However, sixty-two of the sixty-seven volunteers qualified for the St John Ambulance First Aid Certificate.

Each day two of these first-aid specialists attended the RAP to be trained as assistant medical orderlies by Armstrong. Those who were the more capable and more interested were given extra periods in the RAP for instruction and practical work.

The Dodger, who was digging ditches or, rather, leaning on his shovel whilst other mugs did the digging, asked his mate Billy rhetorically, 'Why do some blokes get to bludge in the RAP while we have to shovel mud?'

Having medical orderlies without equipment was like having gunners without guns. Rowley's medical orderlies were unofficial and there was no provision for them to be equipped from the quartermaster's stores. The military machine expected its officers to use their talents for self-help. With the assistance of Radio 2UE, Sydney, an appeal for funds raised several hundred pounds which were used to buy ten steel boxes to contain drugs, dressings and other equipment for troop medical orderlies, and also haversacks, one for each gun, observation post, kitchen, and small groups such as signallers.

The steel boxes were lined with felt and unbleached calico with pockets so that each item could be put in a specific place and readily identified, even in the dark. Each haversack was equipped with shell dressings and tourniquets and became part of gun stores, to be looked after like any other gun store. The gun sergeants, having realised that the troop medical orderlies were first and foremost gunners and would only be called on to act as medical orderlies in emergencies, approved of the whole arrangement and were as proud of the medical haversacks as they were of the rest of their equipment.

The haversacks and steel boxes were, of course, of little use as medical equipment unless they contained medical supplies, and even bottles to contain drugs were in short supply. But Colonel O'Neill had contacts and bottles were supplied through the good favour of Australian Chemical Industries. It only remained to fill them.

Professor Davies' unit was breaking camp. A few days before it was due to leave, Rowley was invited to the final formal mess. By the time the port had gone several rounds, Prof Pete's charming stammer had become more noticeable, an indication that he was well wined and in a very mellow mood.

Rowley, well dined and wined himself, made a steeple of his hands and bowed over it in thought. Prof Pete knew the gesture . . . C.R.B. Richards was evolving a bright idea.

'Pete, that whole idea of troop medical orderlies has turned out very well. I've been thinking . . .'

'Really, my dear boy? So . . . so you have . . . reached . . . a . . . a . . . conclusion and I'm about to . . . to be the victim.'

'Well, if you have any surplus supplies . . . that could disappear . . . I could use them. They could be accounted as having been used in the hospital here, couldn't they?'

'P-pass the port. Can you visualise the p-paperwork that'll land on me. I'll be accounting for those s-supplies, in . . . in triplicate, until the end of the damned war. Go ahead, don't mind me, you . . . you young scrounger.'

Pete waved to the quartermaster to join them and told him that Captain Richards would be calling in the next morning to pick up some stores and would the QM please do what he could to assist. Little did the quartermaster realise what was to happen. Rowley arrived the next morning with a truck, started at one end of the top shelf in the store and worked through all the shelves, methodically assisting about 50 per cent of the remaining supplies to disappear. With contributions obtained by tactful scrounging in other areas, he acquired eleven extra cases of medical stores for his RAP.

By July of 1941 each troop of the 2/15th had two or three proficient orderlies, the chests and haversacks were stocked, and hygiene drill was carried out with as much precision as gun drill. Each troop medical orderly, equipped with a nominal roll of all men in his troop, was stationed in the ablution area every morning to check each man for early signs of skin trouble as he came out of the showers. As the Dodger said, a man couldn't even have a pimple on his buttock without having a dressing slapped on it.

The birds were singing that the Regiment would soon be going to Malaya and a bird sang to Rowley that there were a couple of fellows from Malaya on a visit to Sydney who might be mines of information on tropical medicine: Reg Nutt, a planter from around Kuala Lumpur, and Jim McNiven, a doctor from Ipoh. Rowley went on a brain-picking expedition which provided some tips that weren't in any text books.

The 27 Brigade, which included the 2/15 Field Regiment, received embarkation orders in July 1941. On 15 July Rowley sailed on *Katoomba* to Perth with the main body of the Regiment and transhipped to the Dutch *Sibijak*, en route to Singapore.

3

A Soldier Being a Doctor

The island of Singapore, seen on a large-scale map, has somewhat the shape of hibiscus flowering at the tip of the Malay peninsula. The little island of Blakang Mati at the southernmost point is the calyx from which the petals, in five segments, spread. The blossom is almost bisected by Woodlands Road, from which the one-mile causeway runs across the Johore Strait to the mainland. When Great Britain gained control of the island in 1824 and established it as its bastion of Empire in the Pacific, naval power was supreme. Obviously any attack on Singapore by the great European powers would come from the sea, and there weren't any other powers. China was asleep. The kingdoms and principalities of the Malay peninsula were Asiatic, generally still in the sixteenth century in most aspects of life, and occupied with historical quarrels of their own. The French had gained control of Indo-China. But no Frenchman would dream of fighting through the daunting jungles of Burma, Siam and the Malay States. If France, Germany or Holland should decide to attack Singapore the attack would come from the sea. By 1914 the thinking had not changed, nor had it changed by 1941. The great fortifications of Singapore were all at the southern end of the island, close to the calyx of the hibiscus.

If the Japanese entered the war they might certainly attempt to take Singora, Patani and Kota Bharu—the significant ports on the east coast of Thailand (Siam) and Malaya—but any force attempting to take Singapore from the landward side would have to land at Mersing on the east coast of Johore, sixty miles north of the Malay Strait. An attack on Singapore from Malaya would be doomed to failure, but defence had to be prepared in case the Japanese were militarily inexperienced enough to try it.

During 1941 the British High Command had positioned brigades of the Indian Army, as being most suited for fighting in the tropics, at Kota Bharu on the eastern coast of Malaya just south of the Thai border, and at Jitra and Sungei Patani on the west coast. Should the Japanese land in Thailand and attempt a thrust across the most narrow northern part of the Malay peninsula, the Indians would contain them. Indians were also at Kuantan (22 Indian Brigade) but the HQ for the 9 Indian Division was at Kuala Lumpur. Staff headquarters of the Australian Imperial Force 8 Division was also located at Kuala Lumpur in February 1941, when the first brigade group arrived aboard the *Queen Mary*. The role of the AIF was seen to be that of a mobile reserve.

Mersing, on the Johore east coast, however, was the one possible landing point for an attack on Singapore. And Japan had a portion of its navy on supposedly peaceful manoeuvres in the South China Sea within striking distance, so British High Command revised its thinking and allotted a new role to the AIF. In July when it was known that further units of 8 Division were ready to leave Australia, it decided that the Australians would form a fixed defence force for the State of Johore. Advance Division HQ was moved to Jemaluang, with the 22 Australian Infantry Brigade at Mersing a few miles to the north and the 27 Australian Infantry Brigade a little to the south of Division HQ.

The 27 Australian Brigade had arrived at Singapore on 15 August. Rowley was on deck, sitting on one of his precious steel medical boxes. Jim Armstrong, Sid Brien, Pinky Rhodes and Don Booth—the unofficial troop medical orderlies—were guardian dragons of the other nine chests. The ship was crawling with medical people and, according to the Dodger, if a bloke had to get real sick or have his teeth pulled, that was the time to do it. The 2/10 Field Ambulance and four dental units were also attached to the Brigade.

After disembarkation the 27 Brigade was moved rapidly to its position near Mersing. For the next three months the Australians were digging in. The 2/9 Field Ambulance, 22 Brigade, constructed an Advance Dressing Station and underground operating theatre which would withstand anything but a direct hit. It could accommodate 600 stretcher cases for at least a week, with blast-proof recesses to shelter the patients.

Tragedy struck the 2/15 Field Regiment when the CO, Lieutenant-Colonel John O'Neill, died of injuries received in a motor vehicle accident in the first week of November 1941. Other officers in the vehicle received minor injuries. The popular and highly respected Colonel was buried with full military honours in Malacca and the officers and men of the Regiment each mourned his loss in his own way.

The 2/15 Field Regiment, 27 Brigade, was spread out west of Mersing in the Muar-Segamat area. Rowley set about establishing a system of operation for his RAP and troop aid posts, now a scattered brood.

He needed transport.

The truck was yellow, a dun colour which blended nicely into desert but stood out like an exotic bloom in the jungle. It had newly arrived as part of a fleet of trucks sent from the Middle East so hurriedly that their camouflage paint had not been changed. It had a wistful look to it, as if it wanted to be taken away from the other yellow trucks in the vehicle park and allowed to run in the jungle. It was a neat half-ton job. Rowley loved it. He chuckled when he saw it.

'It ought to be liberated,' he observed to the Transport Officer.

'Don't look at me,' the Transport Officer protested. 'Prof Davies' quartermaster told me how you go about liberating things.'

But Rowley liberated the truck and filled it with the stores liberated from Holdsworthy and other sources. Owing to the scattered positioning of the troops, it would not be possible to evacuate casualties through the 2/15 RAP in the event of hostilities. They would need to be treated first at the troop aid posts he had established and then evacuated through the field ambulance units to the Casualty Clearing Station. Rowley visualised his own role as one of organisation and liaison. He was so busy buzzing around that you couldn't catch him long enough to get him to eat, Sid Brien and Don Booth complained heatedly to Jim Armstrong. Rowley always forgot to eat when he was preoccupied.

The code word for the commencement of hostilities was 'Raffles'. Sir Thomas Stamford Raffles was the English colonial administrator who had the foresight to buy the almost uninhabited island of Singapore from the Sultan of Johore in 1819 and founded a well-fortified port there. His famous name was perpetuated during the last years of the nineteenth century and well into the twentieth in Singapore's leading hotel, where Britons of the colonial service, and planters, drank gin and tonic and danced in an age of gracious living when the sun never set on the British Empire. Throughout the history of Raffles the man, and Raffles the hotel in its heyday, either French, Dutch or Germans had been the enemy.

The code word 'Raffles' was used on 7 December 1941 and the enemy was not as envisaged for more than a century. The Japanese had attacked Pearl Harbor, the American base in Hawaii. The following day they landed troops at Singora and Patani in Thailand and Kota Bharu just south of the Thai border, and commenced their drive to the west and south.

The Japanese force at Singora crossed the peninsula at its narrowest part, on a southward slant, taking every village en route. The force which landed at Patani followed a parallel course some forty miles further south, converging on the area defended by the 11 Indian Division, forcing them into rapid withdrawal. The 9 Indian Division was forced to withdraw from Kota Bharu, and suddenly the 2/10 Australian General Hospital at Malacca was almost in the front line. The AGH commenced a withdrawal in stages to Singapore Island itself. By 10 January, only thirty-three days after the commencement of hostilities, the Australian Imperial Force under its GOC, General Gordon Bennett, had formed a defensive line at Segamat only 116 miles from Johore Bahru—the gateway to the causeway to Singapore. As well as the AIF 27 Brigade Group and the 2/10 Field Ambulance, there were under Gordon

Bennett's command the remains of the 9 Indian Division, a party-trained Indian brigade, the 45th, at Muar, and a Brigade of the British 18 Division. This mixed group, named Westforce, remained on the Segamat line until the remnants of the shattered Indian 11 Division had moved through to the rear. The Japanese were spearing down both the west and east coasts of Malaya, systematically taking every significant village on the map.

At Gemas, some twenty miles north of Segamat, the 2/30 Battalion, AIF, took the first attack of the enemy against an Australian force. They repulsed the enemy inflicting heavy losses (three guns lost for six tanks and mortars) and, blooded for the first time, were jubilant. They had the Monkeys on the run.

Almost no one of the Australian troops had seen a live Japanese soldier at close quarters, but thanks to the steady brainwashing of the propaganda machine they knew the face of the faceless enemy. They had anthropoid features, little, glaring eyes crossed behind enormous spectacles, a prognathous jaw and lips curled back over huge buck teeth. The enemy was barbaric, heathen, unintelligent, very closely related to the monkey. In fact the propaganda machine referred to the Japanese as 'Yellow Monkeys'. As the military machine transformed human enemies into straw dummies so that its components could be taught to rip and twist with bayonets into living flesh, spilling out shining coils of gut, so it was obliged to make the collective enemy always less than human.

Unfortunately the Monkeys knew some intelligent tricks. They skipped around fifty miles behind the Australian line and by 16 January 1942 were attacking the Indian brigade at Muar. It was a bloodbath. In a few days the 45 Indian Brigade had been wiped out. The town of Segamat was demolished to protect Yong Peng and Ayer Hitam, and the Segamat area was abandoned before it also was destroyed.

Stubborn and gallant resistance by remnants of the Australian 2/19 Battalion, 2/29 Battalion, 65 Field Battery and 4 Anti-Tank Regiment held up the Japanese Imperial Guards Division for nearly one week, which enabled the entrapped main body of Westforce to escape to Johore Bahru via Kluang and Ayer Hitam. In retaliation for heavy losses in tanks and men at Bakri and Parit Sulong, the Guards Division committed one of the campaign's worst atrocities—the torture and massacre of helpless wounded soldiers.

Rowley nearly came to grief twice during this battle. On 14 January he visited 65 Battery at Muar to replenish the medical stores of Pinky Rhodes and on the way back was repeatedly machine gunned from the air while on the causeway between Bakri and Parit Sulong.

When news was received that 65 Battery was cut off at Parit Sulong the signals officer, Lieutenant Russ Ewing and Rowley set off to see if they could contact and pick up any wounded. Mid-afternoon on 21 January they arrived at the 2/Loyals area where a counter-attack had been planned for several hours earlier.

All was peaceful, calm and placid until Colonel Thyer, the General Staff Officer from 8 Div HQ, arrived with a message from General Bennett to get on with the counter-attack. The effect was minimal on the weary and apathetic soldiers whose attack was postponed to 1800 hours, then to 0600 on the following day—and never did eventuate.

Russ and Rowley were so furious with the display of apathy that they set off down the causeway towards Parit Sulong. After some distance the ominous silence became so threatening they turned back. They were, in fact, within firing range of the Japanese who amazingly held their fire, probably in the belief that the truck was a decoy.

Landings had occurred north of Mersing and the 22 Brigade, AIF, was forced to withdraw. The entire defence force was being pushed back on to Johore Bahru, with its back to the Strait and the causeway to Singapore. The underground dressing station and fortifications at Mersing never fulfilled the purpose for which they were intended. The war swept past them and southward.

The Monkeys were very highly trained and disciplined monkeys; month after month they had been training twelve hours a day in jungle terrain, crawling on their bellies through the stinking mud of swamps, sleeping under logs that fluoresced with decay, in clothes that grew fungus as they slept. The greatest lesson the Western forces had to learn about them was that they were capable of fanatical devotion to a cause and had an infinite capacity for painstaking preparation.

The Japanese did not give up. Kamikazi was a word not then familiar to the Allied forces, but they were to learn that for Japs death in battle was an honour, capture a disgrace and suicide preferable to defeat. The Imperial Japanese Army accomplished the unthinkable by advancing 635 miles from Singora to the gateway to Singapore, through the worst jungle in the world, in less than two months.

Although everyone hoped the day-to-day tactics of the fighting retreat down the Malay peninsula were clear to the top brass, all was confusion as far as the troops were concerned. The artillery dug in, fired, retreated hastily, and dug in again. Rowley was out all day in his yellow truck, dashing from regimental HQ to division or brigade HQ to find out where the war was at. He checked the position of Battalion HQ, RAPs, and the location of Field Ambulance units. His troop aid posts (TAPs) were functioning smoothly but the troop medical orderlies had to know where the Field Ambulance units were before they could evacuate their wounded. He checked and replenished the supplies held by the troop medical orderlies.

Between 10 January and 15 January he was shuttling around the Segamat-Muar section. Mists were thick in the early morning and the roads were blocked with traffic: military and civilian vehicles ranging from heavy trucks to ox carts; civilian refugees, native and white, on foot, on bicycles and in any available form of wheeled conveyance. The evacuation of Muar became a confused memory of bombs, rain, roads under water and blown bridges.

Jim Armstrong, promoted to sergeant at the end of December, was in charge at the 2/15 RAP while Rowley was on the road. The three best of his TAP orderlies—Pinky Rhodes, Sid Brien and Tom Cooney, later promoted to corporal—were responsible for the battery aid posts of the three batteries. All gun crews of the 2/15th included a trained first aider. They tended their guns, dressed the wounded, arranged stretcher parties and went back to their guns.

Rowley saw himself as a gunner trying to be a doctor in the confusion of battle. He still wore the artillery colour patch only, and had decided against painting the Red Cross on his now camouflaged truck. What was the sense? He suspected the Japs would not refrain from firing on the Red Cross, despite its supposed immunity in battle.

It was all a debacle, of course, but a glorious debacle if one kept one's mind on the acts of heroism performed by individuals in the stress of battle and tried to blank out the sight of young recruits with bewildered faces who wandered, despondent and leaderless, southward along muddy, shell-blasted roads and through the ranked trees of rubber plantations. Some carried their rifles listlessly. Others had lost them or thrown them away. They weren't much use to them anyhow. Rowley had seen recently arrived untrained reinforcements, less than a month in the army, trying to load their rifles by taking rounds one by one from the clip and shoving them into the breech.

On the other hand, he knew of Lieutenant John Ross and Sergeant Bert Tate near Parit Sulong at the end of the Battle of Muar, who supervised the firing of a twenty-five pounder by Gunner Jack Menzies of the 2/15 Field Regiment in an anti-tank role. It stopped the leading tank of a convoy of six tanks at less than fifty yards and, together with others armed with hand grenades and anti-tank rifles, destroyed most of the remainder. This action saved from almost certain annihilation the remnants of the Force commanded by Lieutenant-Colonel Anderson—about 500 men confined in an area of approximately 1000 by 400 yards. For their part in this action Lieutenant Ross was awarded the Military Cross and Sergeant Tate the Distinguished Conduct Medal. Lieutenant-Colonel Anderson was awarded the Victoria Cross for his part in the Battle of Muar. 'A fitting tribute, both to his own prowess and to the valour of his men', was the verdict of General Percival, GOC Malaya Command.

The order was 'no retreat' but the infantry and artillery had no air support. The few Hurricane fighters were unserviceable. They either sat idle on the tarmac in Singapore or crashed on take-off. Fifth-column activity among the Asiatics was strong and there were rumours of Japanese infiltration of the Australian defence line. 'No retreat' was changed to orders to evacuate Johore Bahru and prepare to defend the island—a bazaar with the guns of the proud fortress fixed so that they pointed southward out to sea and could not be turned to fire northward across the island. In any case, the field guns that could have been used were not permitted to fire on the Sultan's palace in Johore Bahru and other buildings being used as observation posts.

The British, Australian and Indian forces were massing along the northern coast of Singapore for the defence of the island.

On one of his trips to the AGH to check the casualties from the 2/15th, Rowley heard that Jim McNiven, the civilian doctor he had looked up in Sydney for information on tropical hygiene, was at the Goodwood Park Hotel, adjacent to Tanglin Barracks, headquarters for the Australian Forces. Deciding to see if he could be of any assistance, Rowley dashed into the bar of the hotel, mud-splashed, dripping with gear . . . tin hat, gas mask, revolver . . . and pulled up with his mouth open. The bar was all soft lights, sweet music, the gleam of bottles and glasses, and there was a group of men and women in evening dress: the men in tuxedos, the women in long, svelte gowns with fashionable boleros and deep cleavages, and gleaming jewels. They were all playing liar dice. Dr Jim McNiven was one of the men in tuxedos.

Good grief, they don't even know there is a war on, Rowley thought, speechless from amazement.

'Richards, my dear fellow, fancy seeing you here. Bit different from Sydney, eh? You do look a trifle dishevelled. Have a drink. Whisky and water, I think, is your brand of poison.'

McNiven signalled to the Malay bartender and the whisky was served with the smooth, quiet, smiling efficiency of peacetime living. I don't believe this, Rowley's mind kept racing as his hand automatically accepted the drink and his head nodded acknowledgement of the introductions that McNiven was performing.

'What are you doing here?' he asked McNiven.

'Staying here. It's quite a good pub, you know. My place up north is untenable just now.'

'I mean, what are you doing here at all? Shouldn't you be evacuating?'

'No boats, old chap. Actually, nowhere to go. The Nips are attacking Java and Sumatra. I don't think a boat would get far in these waters with the Nip navy and air force around. I wouldn't be much use outside. Untrained civilians only get in the way of you professional fellows, eh?'

One of the beautifully groomed women produced a cigarette from a tiny silver case. Part of Rowley's mind noted that it was a small, pastel-coloured Cocktail Sobrani with a gold tip. A jewel gleamed on her finger. As her tuxedoed companion supplied her with a light he grinned at Rowley.

'Have to keep on keeping on, don't we?'

Who is mad? Rowley wondered. Are they mad in here or are we all mad outside and they're the only sane ones left?

'No thank you,' he replied to Jim McNiven's cordial invitation. 'I won't stay to dinner. I have to get back to the war.'

McNiven escorted him to the door. 'How long will it be, do you think, before they take Singapore?'

'Naturally we'll hold the Island, doctor . . .'

'No you won't, doctor . . . Captain. We British called this place a fortress but we haven't kept up with the times. Our guns are useless and the place is nothing more than a big bazaar. You can't defend a bazaar.'

Rowley knew it, but soldiers weren't supposed to admit such things, to civilians or to themselves.

The defence of Singapore lasted for fifteen days, from 31 January to 15 February, and those days were a confusion of pattern bombing of the city docks, repeated air raids that disorganised the city and cut off the water supply that came from Johore Bahru along the causeway that crossed the Malay Strait.

Hospitals were on the front-line, doctors and nurses carrying on in blacked-out wards with the unnatural calm of the highly trained and disciplined. Rowley moved his RAP from Patterson Road to the grounds of Government House, where the sappers were digging slit trenches across the lawns and flower beds and the 'B' Echelon had established a harbour under camouflage nets for the gun tractors. He noted that his liberated truck had clocked two thousand miles since 9 December.

The commanding officer of the 2/4 Casualty Clearing Station, Lieutenant-Colonel Thomas Hamilton, was preparing to move the CCS to the Hotel Cathay. The Australian Army nurses were being evacuated, a move that created depression amongst the staff and patients for whom the presence of the nurses had been a symbol of hope and comfort. The doctors, though commissioned officers, were moving into the wards to sleep beside their patients, in the hope of providing mental comfort and moral support.

Rowley's medical orderlies were as calm and efficient first aiders as they were fighting men; a vindication, Rowley felt, of the hours of training he had instituted at Holdsworthy, despite protests.

The Australian 8 Division defended the north-west section of the island—the petal of the hibiscus flower immediately to the west of Woodlands Road—with the 22 Brigade stationed on the coast and the 27 Brigade across Woodlands Road which led to the mainland causeway. Indian brigades held the centermost petal and the British brigades were ranged along the north-eastern coastline. The petals of the hibiscus blossomed into scattered flowers of flame as whole sections of the city took fire under the heavy bombardment. Civilian casualties were many. Children and domestic buffalo wandered unattended through the shambles of the bazaar.

As the defenders of Singapore retreated street by street into the wealthy Tanglin Hill residential area, Rowley found the reality of battle assuming an unreal, nightmare quality. In search of a defence post he ran in through the open French windows of a businessman's mansion and found the family at dinner. Despite the shelling that had cratered their garden they had been carrying on as usual. The table was set with damask cloth and candles, silver and crystal, and dinner had been served. But the seated diners were dead, unmarked, but killed by the implosion of a shell blast. As Rowley ran the length of the dining room, the vibration caused a corpse to topple.

He did not stop. There was nothing he could do. The drawing room was a shambles. The shell had blasted one wall. He ran across churned earth and the rubble of a garden wall to the next house. It was intact and unoccupied. An open wall safe and open dressing table drawers in a feminine bedroom testified that the owners had lingered only to gather their jewellery and any other portable valuables. The house would do as a gun post.

During the day of 13 February there were twenty-seven bomber raids, and the raids kept up the next day. At one stage in the morning Rowley heard himself reading the burial service for two casualties of the 2/15th and part of his mind wondered how he came to be doing that job, while on the surface he concentrated on the words of the service. It was strange that his mind functioned on two levels, one of which seemed to be watching him from a distance. It watched him hugging earth close to a tree trunk, with his driver, Vic Carter, and Signaller Teddy Andrews, while Jap planes, engines howling, made their run. The base of the tree was spattered with shell fragments and Vic Carter was swearing because his mess tin had been holed. Teddy Andrews had been hit, blood from a shredded calf soaking the khaki cloth of his trouser leg. Rowley patched him up.

The CO of the 2/15th after the death of John O'Neill, Lieutenant-Colonel John Wright had been wounded in the knee and blown by the bomb blast right into a slit trench. But 'Gentleman John' Wright, a courteous soldier of the old school, stayed in the RAP only long enough for Rowley to suture the knee. Although slightly concussed he refused Rowley's request to be allowed to send him to the CCS.

Then, somehow, Rowley was riding pillion on a motor cycle driven by a crazy despatch rider on a round of treating the wounded. The detached part of his mind told him he was likely to be killed riding pillion, which would be a dreadful anti-climax, but the despatch rider wouldn't leave him. He said he had orders to take Captain Richards to the 2 I/C, Major John Workman and he dropped Rowley off at a sandbagged dugout in the ruined basement of a building that had been the Japanese consulate.

It was blessedly quiet in the dugout after the turmoil outside. Workman was there, and so was Ted Dahl, the Light Aid Detachment Engineer, along with Stuart Ward, the Adjutant, and Battery Commander Frank Ball. They were discussing capitulation and evacuation.

Workman nodded to Rowley. 'This isn't defeatist talk, Rowley. Capitulation is certain. It's only a matter of when. Gordon Bennett has already said he will try to escape. Most of the senior officers believe organised escape for the troops will be too hazardous. What we do is going to be up to groups and individuals.'

'We envisage another Dunkirk,' Frank Ball chimed in. 'There won't be any official evacuation transport. We must commandeer little ships if we can. All we can do is try to get as many of the regiment off as possible.'

'You had survey regiment training in the militia. Can you navigate?' Workman asked Rowley, who nodded. 'Right, then. Try and get hold of some useful maps.'

Rowley thought of the big houses on Tanglin Hill. There had been a well-stocked library. He should be able to find atlases.

He realised that two more men, Lieutenants John Bull and Roger Martin from Frank Ball's battery, had come into the dugout and were listening to the conversation. Then Frank Ball was talking to them, telling them to scout the waterfront for suitable ships.

The quiet-voiced planning seemed as unreal to Rowley as the rest of the day had been. It didn't seem possible they were organising a Dunkirk-style evacuation when their war wasn't much more than two months old. He thought he must be suffering from shock after the near miss by those Jap planes and felt as dazed as the poor old CO had looked after he was blown into that dugout.

Later, after he had found atlases and torn out maps he thought would be relevant, he heard from Frank Ball that Bull and Martin had found two small, coal-burning island traders. On one the engines were playing up, but the other, the *Hong Fat*, was in good condition. Lieutenant Bull was in control and had been ordered to wait off a rendezvous point, a pier in Keppel Harbour. One of Frank's men, Lieutenant Max Carment, was provisioning the ship. Word would be filtered through the regiment as to the *Hong Fat*'s location. After surrender it would be every man's duty to escape if he could. The plan seemed to be working satisfactorily.

By Sunday morning, 15 February, the AIF had withdrawn into a small racket-shaped perimeter surrounding the Tanglin Barracks, about three miles from the coast and on the outskirts of the city of Singapore. Some 5,000 AIF troops in the Tanglin sector were served by an improvised field hospital with accommodation for fifty stretchers and staffed by field ambulance personnel. The 2/10 AGH was installed in the Hotel Cathay and the ambulances were at the cathedral.

The officers of the Australian Army Medical Corps realised with general bitterness that there was to be no evacuation, not even of the hospital patients. Although the Director-General of Medical Services Office at Army Headquarters, Melbourne, had met all requests for supplies promptly throughout the Malayan Campaign, only 118 sick had been sent away in the peacetime passenger liner, *Orion*, on 31 December, and 127 sick and wounded were evacuated on 10 February aboard the *Wah Sui*, a small coastal vessel fitted out as an inter-port hospital ship by the Deputy-Director of Medical Services, Malaya Command. But no hospital ship had been sent from home and defence and evacuation were equally impossible. The castle had swept down the chess board, taking pawns and knights and bishops and queen, and those pieces were to be taken off the board and laid aside until the game was over.

The artillery fire ceased at 1130 hours on 15 February but the Jap planes continued bombing the island, particularly Keppel Harbour which was a chaos of small shipping. John Bull on the *Hong Fat* came under fire from a Jap machine gun and one of his men was killed. He moved to a safer position about 800 yards off the pier and waited for his boatland of refugees. Men drifted down in twos and threes, no one quite sure what was happening. All Singapore was in chaos.

That evening Rowley called to see the ADMS, Colonel Alfred Derham, to receive the daily report and orders. As he waited outside the entrance, the GOC's staff car screeched to a halt and General Gordon Bennett swept into Division HQ with Major Charles Moses and another staff officer.

This has to be it, Rowley thought. Waiting, he could hear muffled voices but could not distinguish words. Then there was a sound like marbles being thrown into a tin dish. Later he realised that he had heard the revolver bullets and other ammunition being collected. The whole visit was over in less than half an hour and the GOC and his officers had been driven off at high speed, farewelled by the stunned officers of Division HQ.

Colonel Derham saw Rowley, came over, and put a hand on his shoulder.

'Richards, the GOC has just advised us that General Percival has capitulated unconditionally to General Yamashita. The Malayan Campaign has ended . . . it was a tough campaign, although short . . . but we may have a much tougher and very much longer and different one ahead.

'The GOC has just announced that he has relinquished command in favour of Brigadier Cecil Callaghan, Commander Royal Artillery. He intends to attempt to return to advise Australia of the details of the campaign. He has ordered that all officers will remain with their troops. There is to be no organised evacuation or attempt to escape. If escapees are caught the penalty will be death.'

The detached part of Rowley's mind returned to the scene in the dugout, the planning of a Dunkirk. That's the end of the voyage of the *Hong Fat*, he thought. The other part of his mind noted that Colonel Derham was paying tribute to the work of the medical corps and thanking him for his contribution.

Then he was on his way to RHQ, where he discovered that John Wright had already been called to Division HQ. He decided he should not talk about the scene he had witnessed until after Wright had returned. It was for the CO to tell the men officially that the war was ended—and they were prisoners, even ordered not to escape.

But no one then could contact John Bull. The bewildered lieutenant had some fifty men aboard the *Hong Fat* and, uncertain what to do, he had gone to the pier and reported to the British officer in charge of evacuation, Commander Alexander, Royal Navy.

'I must clear craft out of the harbour,' Alexander told him. 'Leave while you can. Now. And good luck to you.'

'But I haven't a full load, and I can't contact my commanding officer,' Bull protested.

'Communication is impossible. Get away with the men you have. I don't want you in the harbour,' Alexander replied.

Reluctantly, Bull headed the *Hong Fat* out into the stream, probably the only evacuation ship to take AIF troops out of Singapore on the day of capitulation. He had no way of knowing that there had been a change of orders and the *Hong Fat* headed towards British minefields and an epic of her own.

Meanwhile, the bone-shaking bombardment had ceased, and Rowley discovered that the respite was almost worse than the noise. At least the noise made any sort of thought impossible, other than the automatic thought necessary to carry out the automatic actions of a well-drilled fighting component. It was just as well. Independent thought might include a realisation of one's predicament as a living organism in danger of being blasted into bloody fragments at any moment. Rowley climbed into his truck and sat with his hands limp on the steering wheel, feeling all the energy that had sustained him through days of bomb, shell, mortar and machine-gun fire drain away. His legs and arms were suddenly weak, aching. The carousel music in his head played, 'It's over, it's over, it's over. Thank God.' It was Sunday evening, 15 February 1942 and his subconscious tossed up the information that it had been a Sunday evening, on 3 September 1939, that he had heard the declaration of war given by the voice of Neville Chamberlain.

He looked out the side window of the truck onto the placid green Tanglin Golf Course. Gunners, their guns silent, were sitting on heaps of debris and chunks of masonry; one chap, improbably, on an office swivel chair. Some rested their heads in their hands, some stared blankly ahead letting their hands dangle between their knees. It was over. He didn't have to do anything about the war anymore. Tomorrow there would be the problems of something else; quite what it was he couldn't think. But he didn't have to think for just a little while. It was all up to somebody else.

Monday morning was quiet and peaceful, strangely like the peace of a Sunday morning at home. The men of the Royal Australian Artillery slept, checked over their kits, visited other gun posts, other batteries of their regiment, checking which of their mates were still around and which were listed amongst the killed or missing. Without enthusiasm they drank champagne from the Tanglin Club and from the stocks in the homes of the wealthy that stood deserted on exclusive Tanglin Hill or were occupied only by the dead. There were no owners to drink it and the soldiers felt that they deserved it. A long time might pass before they tasted alcohol again. They discussed the actions of General Gordon Bennett.

The furphies were many. Some said bitterly that he had left before capitulation, but most agreed that he had contrived to escape from Singapore, hoping to make his way back to Australia to continue in his duty as a fighting man. Some said he was right to go; some that he should have stayed with his men. Wherever he heard the furphies, Rowley repeated how he had been at Div HQ when the General announced the surrender. But he knew the arguments would go on. Recrimination was the burden every leader had to pay for decision-making.

At the 13 Australian General Hospital in St Patrick's School, Lieutenant-Colonel Thomas Hamilton was dressing a wound when a patient warned that there was 'one of them little Monkeys' behind him. He turned and for the first time saw at close quarters a Nipponese infantryman, 'strangely like an anthropoid ape . . . dressed in peaked cap and dirty khaki drill breeches, with puttees and black rubber shoes'. The Nip left rapidly when Colonel Hamilton waved a pus-soaked dressing under his nose.

General Yamashita, the 'Tiger of Malaya', paid a personal call at the Cathay Hotel where the 10 AGH was established, arriving in a red-flagged staff car and showing courtesy to the medical officers.

The Alexandra Hospital had not been as fortunate as the establishments at the Cathay and at St Patrick's, where the officers had been careful to maintain the role of the medical corps and had placed lighted hurricane lanterns on the Red Cross sign. Trigger happy Indian Sepoys had fired on Japanese troops from the Alexandra, attracting the enemy to the wards which they went through, indiscriminately bayoneting the sick and their attendants. A Japanese officer in white gloves called at the Alexandra during Monday to apologise for the massacre. There were instances of bayoneting and executions but, generally, General Yamashita maintained strict discipline and forbade alcohol, so the vanquished were spared the frenzy of rape and massacre—that terrible disaster of war for the vanquished when the victors are inflamed by drink and the heady stimulant of conquest.

It was only natural, however, that the conquered should hate and despise the conquerors. The rhythmic thud of marching feet as the victors marched through the city was a sound that left permanent scars in mind and spirit. A mingling of hatred and pride carried the defenders of Singapore through their last defiant, grand gestures as a fighting force.

Japanese orders were that all military and white civilian prisoners were to be incarcerated at Changi Barracks, on the extreme eastern point of Singapore Island, and in the camps of Selarang and Birdwood.

All patients and medical units were to be moved within three days and only a limited number of vehicles were allowed. Fractures had to be resplinted for transport; the lives of transfusion cases would be jeopardised; medical supplies had to be salvaged and protected from confiscation by the enemy; contact had to be made with the International Red Cross before the prisoners were cut off from the outside world.

Despite his leg wound, Lieutenant-Colonel Wright elected to march into Changi at the head of his men and limped the sixteen miles of the route through the devastated city. Rowley finally wore a Medical Corps colour patch, brown with a grey surround, with a small Artillery patch superimposed on his hat and had a Red Cross sign painted on his truck. He considered the order limiting the number of vehicles permitted for the move and rejected it. 'I'm not leaving our supplies for the Nips,' he said, and with John Wright's knowledge he took his truck to Changi, driving it himself since extra vehicles were certain to bring down the wrath of the Japs.

The Japanese appeared to be moody, emotional people who could change instantly from almost-friendliness to unbridled rage if disobeyed. There were many stories around already about encounters with the enemy through the long night after capitulation. Although those within the boundaries of the hospital locations and protected by the sign of the Red Cross had been unharmed, except for the Alexandra massacre, trucks attempting to leave those sanctuaries had been stopped, their drivers and passengers fired on, even bound and threatened with execution. There had been executions, captives kneeling in ditches and shot through the back of the neck in the last hours, minutes even, before capitulation. The thought of execution in that manner turned Rowley cold. It made him think of the slaughtering of steers at the abattoirs outside Kyogle: the beast standing in the race unable to move but scenting death, eyes rolling in panic, a flick of foam about the nostrils; the stockman taking aim, the sharp crack and brief whine, and the beast collapsing silently, its knees folding. But steers were shot between the eyes, not in the back of the neck. Men, generally, died with greater control of the outward symptoms of fear but were they any less afraid? Rowley preferred not to be executed but he was stubbornly possessive of his medical supplies.

'Hey Doc. Y'need a driver, don't yer? I'll volunteer.'

Poised on the running board of his truck, Rowley looked over his shoulder. He saw a young fellow, big, well-fleshed and good-looking if the flabbiness of self-indulgence in his face was unnoticed, and the closeness of the eyes to the bridge of his nose. It was Dodger. Of course he would have survived. The Dodger would make sure that his was not one of the bodies that was left by the roadside, or half-buried in the rubble of bombed masonry.

'Don't you know how to address an officer?' Rowley snapped.

The Dodger looked astonished. 'But it's all over,' he protested. He had been drinking.

'You're still a soldier, Gunner. Pull yourself together and act like one,' Rowley told him crisply. 'Report back to your troop commander like the rest. It'll help you to dry out.'

The Dodger straightened out of his slouch and made a sketchy salute. 'Yessir. I only wanted to help, sir. Doesn't seem right for an officer not to have a driver, sir.'

Rowley had no intention of explaining why he did not have a driver, or want one. That actions, even disobedience of orders, were sometimes justified by the ends they served would not be within the Dodger's understanding. In that he would only find an excuse for licence.

The salute pulled back the Dodger's shirt cuff far enough for Rowley to see the glint of two gold wrist watches on his arm. And there was a ring flashing white fire on the little finger of his thick hand.

'You've been looting,' Rowley accused.

'The Japs are doing it to our blokes,' the Dodger whined. 'Only getting a bit of our own back, sir. The Monkeys I got these from don't have no use for them no more.'

'You're a vulture,' Rowley told him. 'What about the diamond? That wasn't on a Jap. You got that from a Chinese merchant. Or was it the corpse of a white woman?'

The Dodger's face dissolved into folds of self-pity.

'They wasn't needing it,' he whined. 'Seemed a waste. That posh house, silver and candlesticks and fancy evening clobber, but not one of 'em alive. The other blokes took things too. Didn't belong to anyone anymore.'

With appalling clarity Rowley saw once more the home he had entered on Tanglin Hill, where the family sat dead at dinner. Had it been that home, or another like it that Dodger and his mates looted? Dodger would not have been gentle or respectful as he took the ring from a dead hand. Did the corpse fall as he robbed it?

Sudden fury caused him to choke over his order. 'Report with that stuff to your troop commander. I'll check that you do.' His glare must have been ferocious for Dodger seemed to shrink and pull his head into his shoulders as he scuttled away.

Without waiting to watch the Dodger's retreat, he climbed into the truck and revved the engine angrily. Where the hell was the glory and the pride? Where was the black of the baddies and the white of the goodies—or was there just a murky greyness that swallowed all the ideals of God, King and Country, of Home and Family? He realised, as never before, the importance of discipline, as imposed by the military machine and as imposed by the self upon the self. If discipline broke down, if morale broke down, if men became scavenging animals so quickly, how were they all going to survive? The old ideals weren't cliches, they were an integral part of survival.

Well, Rowley thought wryly, I've posed a problem for Dodger's troop commander. What's he to do with the stuff? The owners are dead and there are no civil authorities to take responsibility for valuables. Oh Hell!

He wrenched the gear stick and let the truck crawl into the slowly moving procession that wound via Napier Road, Tanglin Road and Newton Circus to Changi Road and imprisonment. The time was 1530 hours and the route to Changi covered sixteen miles.

That day John Wright marching with a limp at the head of his men into Changi Barracks embodied all Rowley's ideals. The dignity of the officer was a salve to the smarting abrasion of Dodger's jackal behaviour. Then, on arrival at Changi, Gentleman John was berated before his men by a junior staff officer from 8 Division Headquarters for bringing more trucks loaded with rations into the prison than was officially permitted . . . although others were doing likewise.

Changi, despite its barracks layout, was now prison for soldiers and civilians alike. Those who had done their duty bravely as fighting men, those who had lived law-abiding and useful lives as civil servants, clerks and traders, found themselves incarcerated as if they were common criminals.

Morale was at its lowest among those who had no duties to perform, whose lives were suddenly bereft of all purpose. For those who had His Majesty's commission it was, perhaps, easier, for they still had a job to do. And there were others, both soldiers and civilians, who found jobs to do because it was in their nature to fight and not to yield.

The night of Rowley's arrival at Birdwood Camp in Changi was a busy one. Despite the disgraceful exhibition of a junior staff officer castigating Colonel Wright publicly, High Command was glad to have the rations and the extra medical supplies that had been smuggled into the prison by resourceful medical officers and was quick to issue orders for their safeguarding. The gunners of the 2/15th purloined, scavenged and acquired from bombed buildings the materials to build walls within walls, and before morning Rowley's orderlies had removed all the medical supplies from his truck and stacked them behind false walls, along with the liberated rations.

The staff of the general hospitals in Selarang Barracks were taking similar precautions and were ingenious in devising temporary concealment. Whitewalled offices which had been turned into dormitories had picture rails of black wood, and along these rails, a couple of feet above head height of the Japanese, the black cylindrical containers of Atebrin tablets—the prophylactic treatment for malaria—were laid end to end. They were not detected during snap inspections by the captors and vindicated the theory that the most obvious hiding place is often the safest. Later, on orders from their own headquarters, the Australians were to bury caches of drugs and medical equipment when it became obvious that various nationalities incarcerated in Changi were not loath to engage in unrequested borrowing or commandeering from their allies. Among those who buried the supplies were several of Rowley's men.

Rowley made his own reconnaisance of the prison camp as the fox familiarises itself with new territory before digging its earth. Selarang Barracks, adjacent to the Birdwood Camp, occupied the most north-easterly promontory of the island. Its huge barracks buildings and its officers' quarters, both flats and self-contained homes, spread over acres set with parade grounds, lawns, groves of trees and tropical shrubs, and two swimming pools. Within sight was the blue ribbon of the Strait of Johore.

As well as the swimming pools, tennis courts and a cricket pitch had been provided as amenities for HM armed forces. But palatial it may have been for 2,000 men, it was now packed with thirty times that number of prisoners of war: British, Australian, Indian and Dutch. In one married-quarters building over 200 men were occupying space designed for six families. Most of the beds had disappeared from the barracks and the men had to lie on floors of concrete or wood with their kit bags for pillows. Already some of the enterprising were making camp cots from salvaged wood or bamboo, of which there were thick groves on the slopes above the beach.

The accommodation was appallingly overcrowded, Rowley thought, but the living conditions weren't intolerable. The men could get exercise, fresh air and would even, depending on the whims of their captors, be able to swim. He passed a noisy group on a lawn playing two-up. Some were lounging against barracks walls playing cards, while others were simply lying down and staring at the sky, waiting for someone else to take the initiative, tell them what to do. Well, ways to keep them occupied would soon be organised.

Hygiene was going to be a major problem. Ablution and latrine facilities were inadequate for such a vast number of men, as were the kitchens. Rowley remembered the long talks with Professor Pete and John Loewenthal, the course at the School of Tropical Medicine, the talks with Dr Jim McNiven and Reg Nutt, and the visits he had made to the medical establishments in Kuala Lumpur before 'Raffles'. He and all members of the Australian Army Medical Corps were going to have work enough keeping the men healthy until the war ended or there was an exchange of prisoners, either of which could be weeks or even several months away.

He stopped suddenly, his attention caught by a group of three sitting in the angle of a wall, engaged in opening a tin of condensed milk with a clasp knife. Although the kitchens were not yet functional and the morning ration had been no more than tea and a handful of hard biscuits, these fellows, like many others, had been provident in carrying a few tins of food in their kit bags or haversacks. An empty tin with a bright fruit salad label lay beside their dixies. Its contents had been portioned out amongst them as a topping for the dry biscuits and the resultant mush in the dixies was about to be given a dash of condensed milk. It was not the food that caught Rowley's attention, but the small black flies that crawled unheeded around the rims of the dixies and over a tin spoon left carelessly on top of a haversack.

'Get those dirty damned flies away,' Rowley exploded. 'What do you think they've been crawling on? There are more than open sewers in Singapore. There're corpses that haven't been buried yet. Share your food with the flies and they'll do for you what the Jap shells didn't.'

The three soldiers stared at him open-mouthed. At least they were none of them his men. Some other RMO would have them on sick parade with 'the runs'.

'Unless you relish an almighty bellyache, you'll not only keep the flies off, you'll dip those dixies and spoons in boiling water before you eat . . . every time you eat . . . and wash your hands. Remember it. Faeces, food, fingers and flies spread dysentery.'

He marched off towards his temporary RAP, thinking that enforcement of hygiene was going to be the biggest factor in ensuring the survival of the men as prisoners of war. The other factor was morale-boosting.

'Good Ghost, what was that?' one of the castigated soldiers exclaimed.

A soldier who had been lying nearby with his hat over his face, apparently asleep and with his bare feet resting on a kitbag so that the breeze could cool his blisters, lifted the hat from his face and said: 'That, my friends, is the bloody RMO of the bloody 2/15 Field Regiment. Name of Captain Richards and top of the shark list from way back. You should have a load of the drill he put us through every morning ever since my draft went into camp and right up till the fighting started. Bloody hygiene parade. And that same bastard drives into Changi in comfort while blokes like us march the whole sixteen miles taking the skin off our feet.'

The Dodger eyed the partly opened tin of condensed milk thoughtfully.

'I'll toss you for that tin,' he offered. 'Two quid if I lose.'

Dodger was flush with money. There had been a wad in a desk drawer in the house from which he had taken the diamond ring. Dodger didn't lose. His special pennies were cherished, and carefully concealed possessions.

Rowley's RAP was established at Birdwood Camp, Changi, and from 18 February until 9 March 1942 he reintroduced the skin inspection parades after morning showers and continued to instil into his men the need for strict cleanliness of eating utensils. He wrote a concise report on the medical arrangements within the Regiment since its formation, and the treatment and evacuation of casualties during the campaign, listing the nature and number of casualties:

> 1. BATTLE CASUALTIES. Killed in action, 19; missing believed killed, 11; wounded in action, 59; died of the wounds sustained in action, 4. Seventy-five percent of these casualties were suffered in 65 Battery, which engaged in the Muar campaign.

Rowley, a dedicated compiler of statistics and graphs, continued on with Part C (2) of his report:

2. NON-BATTLE CASUALTIES. The total of 133 'sick' casualties evacuated from the unit during the period 8 December 41–14 February '42 were divided thus:

Malaria	—	52	
Dengue	—	24	
Dysentery	—	9	
Injury	—	17	
Tinea	—	3	
Others	—	28	TOTAL 133

3. SICK WASTAGE. The weekly sick wastage expressed in evacuations from the unit is shown as a percentage of an average of 600 personnel in the Regiment. For comparison, the average weekly sick wastage for the four weeks prior to 8 December is shown. For the sake of completeness, the weekly wastage from battle casualties and missing is also shown:

Percentage for week ending:

	Pre. Dec. 8	Dec. 13	Dec. 20	Dec. 27	Jan. 3	Jan. 10	Jan. 17	Jan. 24	Jan. 31	Feb. 7	Feb. 14	Average weekly %
Non-battle casualties	0.08	0.5	1.2	1.3	1.7	2.2	2.2	2.2	4.3	3.7	2.2	2.9
Battle casualties							0.8	9.7	2.0	0.5	2.5	3.1
Missing								6.7		0.3	7.2	3.3

PART D. REMARKS

The figures shown in C demonstrate the following:

i Malaria and Dengue responsible for 57% of sick wastage, i.e. 76.

ii Only 3 cases evacuated with tinea. Due largely to the excellence of attention by TAP personnel.

iii Only 4 cases of wounded in action who subsequently died. Due to the efficiency of first aid treatment by TAP personnel and rapid evacuation to and through medical units.

iv Excluding malaria and dengue, the average weekly sick wastage was 1.2%.

v Peak of incidence of malaria was from 25 Jan. to 7 Feb.

vi The average weekly sick wastage of 2.9% was slightly lower than the average weekly battle casualty of 3.1%.

vii The training of TAP personnel and establishment of TAPs was amply justified as demonstrated in the prevention of sick wastage and in rapid adequate first aid treatment and evacuation.

He read the report through and signed it—'CRB Richards, Capt AAMC, RMO att 2/15 FD REGT RAA AIF'.

Yes, it was the yield of the exchange of ideas among people like Professor Davies, John Loewenthal and other like-minded soldier doctors. The army trained men to kill, to fight, to hate a human enemy. It had to be so when an aggressor threatened the way of life of a nation, the liberty of a people. But soldiers also had to be taught to survive by staying healthy, by learning routines for the treatment of the wounded.

And what happened when the raison d'etre of the soldier was taken away? The defeated, those in the Limbo of the Inferno who had not died in battle nor escaped to fight on, those who were in the uncertain purgatory of prisoners of war, how were they to survive? Historically, boredom and its attendant depression, the skin and bowel complaints that arose when large numbers of people were confined together, the vitamin deficiency diseases like pellagra, were the problems that confronted medical officers. Historically, of course, it meant the history of wars in Europe. But what other enemies would have to be faced on this small island just north of the equator, at the tip of a peninsula spearing downward into the Indian Ocean on the west and the South China Sea on the east?

Here was a climate inimical to man, and to white man in particular. For a century and more the white *tuan besar* of trade and civil service had succumbed, despite all the amenities their governments could provide, to malaria, boredom, the homosexuality in public servants or promiscuity of men cut off for long periods from women of their own race, alcoholism, and all the neurotic complaints of the expatriate.

How much worse was it going to be for men confined like criminals, not even knowing the term of their sentence? Rowley sighed. The Japs, he realised, would not be the only enemy. Within Changi fighting men of the British, Indian, Dutch and Australian forces were imprisoned. Although there was some social intermingling amongst them, militarily they remained as separate national groups, each with their own area of the camp and their own strata of command.

The top level of command of the Australian POW force rapidly organised itself to combat the two immediate problems of food supplies and boredom.

Australian fighting men, accustomed all their lives to a high-protein diet and generally to three square meals a day, found the abrupt transition to an Oriental diet consisting mainly of rice, some vegetables and dried fish but very little meat, a shock both physically and psychologically. Stomachs and bowels and minds resented the prison rations. The thoughts of the ORs seemed to dwell mainly on food and ingenious means were devised to introduce variety into the boring diet. Men who had never prepared a meal for themselves became inventive chefs, concocting rice cakes from the watery boiled rice doled into their dixies at the kitchen: savoury rice cakes with dried fish and curry powder if some far-sighted mate had brought in a tin in his haversack; sweet rice cakes made with a dollop of jam from a prized tin or with condensed milk, and cooked over a camp fire on a hotplate made from scrounged scrap metal.

The kitchen staff, criticised and maligned as cooks since the earliest of shearers' cooks traditionally had been wherever Australian men were fed en masse, stretched their prowess to make the rations more palatable. Command, with the indefatigable Mr Murchison of the International Red Cross liaising with the Japanese, sought for further increase in quality and quantity of the rations. The daily ration per man was 50 grams of meat, 500 of rice, 50 of flour, 100 of vegetables, 15 of milk, 20 of sugar, and 5 each of salt, tea and oil; the cigarette ration was 40 per month.

Like many others in the camp, young Private Roy Whitecross of 8 Div. HQ began keeping a diary to fill in time. He noted that he was glad he had never taken up smoking as the smokers missed their cigarettes more than a sufficiency of food. Roy missed his three squares and had trouble with his denture which was hurting his mouth and kept slipping; the dental officer unsympathetically told him he was lucky to have a denture at all when so many men having extractions could not be supplied with them.

Rowley calculated that the ration yielded 2,000 calories per day per man with perhaps 500 calories from the 'extras' they were able to supply from their personal hoards or by purchase from the natives. That was at least 500 calories less than a grown man needed, but it was supportable. His sick parades showed an increasing number of men suffering symptoms suggestive of salt deficiency, which he found an immediate cause for concern.

There was an abundant source of salt within a few hundred yards of Changi Barracks—the ocean. The Swiss Family Robinson, Rowley remembered from boyhood reading, had obtained salt from rock hollows from which the sea water had evaporated. Perhaps it would be possible to supplement the salt ration by boiling rice in sea water. The experiment was tried, both decreasing the salt deficiency and improving the palatability of the bland, insipid, previously unsalted rice. It didn't taste like rice as served with curry at home, but it was an improvement, and the saltwater details were popular as the water carriers were able to go swimming from the beach.

Unfortunately, however, a group of water carriers attempted to carry out an escape and the Japanese demonstrated convincingly that they did not make empty threats when they warned that escapers would be shot. Thereafter the water details were not permitted and the bland, unsalted rice tasted worse after a brief experience of that boiled in sea water.

Concert parties, hobby groups and Changi POW University, under Brigadier Harold Taylor, were organised to combat boredom. Colonel White organised a Post Graduate Medical School at which the RMOs attended lectures. Captain Adrian Curlewis, Dean of the Law School, lectured on the Geneva Convention and the rights of prisoners of war. Private Roy Whitecross enrolled for a course in engineering which he found interesting; a vegetable garden was established; Sergeant Jim Lynch, a lawyer in civilian life, was one of the speakers who gave lectures on their specialities in a series of talks organised by Brigadier Taylor. The concert parties, given by moonlight, managed performances with 'Ave Maria', 'Moonlight Sonata' and 'Liebestraum' on the program. For organisers, performers and those who sought mental stimulation, Changi University, the lectures and the concerts were rewarding. Others drifted from one diversion to another, killing time until the day (which they reckoned could not be far distant), when they would be liberated by advancing Allied forces, or repatriated home in a prisoner-of-war exchange.

News received on the forbidden, concealed radio receivers was circulated by word of mouth, furphies multiplied, and it was generally believed that the term of imprisonment would be short.

Collecting, retaining and speculating over furphies was a survival mechanism. Everyone knew someone who knew someone who had access to information heard over one of the hidden receivers. Even the wildest furphies were given serious consideration around the camp fires, for the greatest morale-builder of all was hope of an early release. Had it been generally believed that incarceration would last for years, then the whole planned structure of morale maintenance, the concerts, study courses and debates would have fallen apart before they had been properly established. And with them would have gone group identity and pride, then very quickly personal pride with a consequent disastrous deterioration in social interaction and hygiene. Such hope, however, was to prove only disappointment deferred or reality rejected.

Considering the situation as an exercise in large group psychology, Rowley could accept, intellectually, an ultimate breakdown of social behaviour with lack of group and personal cleanliness so severe that epidemic disease could decimate the inmates of an overcrowded prison like Changi.

Then, in a textbook Freudian situation, the primitive Id, the animal identification of self with its drive for self-survival, became dominant and in the resulting chaos human beings preyed on each other like jungle animals. But he did not believe that any group of civilised human beings could reach that ultimate degeneration. Conditions in a POW camp in the twentieth century could never become so bad. The whole idea was emotionally and intellectually repugnant and unacceptable.

The Dodger and his mates drifted along to some lectures, found them boring, and preferred the two-up games which attracted hundreds of devotees until Command discovered that the men were willing to bet their rations on the fall of two pennies and forbade the pastime. The drop in morale was so severe that Command had to reconsider and withdraw the veto, recognising the need for levels of interest that ranged form 'Moonlight Sonata' to the fall of a coin.

Altogether, life in Changi during the first two months after the fall of Singapore was tolerable, no worse than any of the German Stalags. The 2/15 and 2/10 Field Regiments, with six other battalions, ordinance store, and the 22 Brigade HQ moved into Selarang Barracks in March; 700 men in accommodation designed for a maximum of 200. Beds were constructed in tiers and very soon everyone discovered bed bugs in their charpoys and bug-catching parties became a daily event. Bug catchers became experts on the sexual, family and social life of bugs.

Rowley shared an officers' barracks room, twenty feet by twelve, with Battery Commander Frank Ball, Des Makepeace and Stuart Ward. He established his RAP in a converted orderly room, with the help of his medical orderly sergeant, Jim Armstrong, and settled into a routine: Reveille was at 0800 hours, sick parade 0830 hours, breakfast 0930 hours, work parade 1030 to 1330 hours, lunch 1330 hours, work parade again from 1430 to 1800 hours, tea at 1830 hours, and lights out at 2100 hours. The two and a half hours from tea until lights out were a time for attending concerts, exchanging ideas with his friends (and where Rowley delighted in what he termed 'brain milking'), and studying for Colonel White's Changi University Post Graduate Medical School course. It wasn't exactly just like home, but under the cirumstances it was a fairly sane sort of existence. It would last out their imprisonment.

On 5 March two cases of dysentery reported on sick parade. From his limited supplies, Rowley was prescribing bismuth compound in tablet or powder form 'for unsettled stomachs', magnesium sulphate, followed one hour later by Mist Astringens 'for diarrhoea'. The two dysentery cases were treated with emetine. For every man of the 2/15th Rowley had a prophylactic prescription for dysentery: 'There's a drum of boiling water outside the kitchens. Dip your plates and eating tools into it before you get your meals.' He noted, with justifiable pride for the RMO nicknamed the Baby Doctor, that other units had up to 50 per cent of their sick parades reporting with dysentery.

The dysentery epidemic commenced on 17 March, peaked on 27 March and was on the wane by 31 March. By 22 April the epidemic was under control with a sick parade of thirty. There had been 238 cases out of the 700 men in Selarang Barrack: 110 mild diarrhoea, 34 mild dysentery and 94 true severe dysentery. Preventive hygiene was still winning, and his men were cooperating.

The medical problems of prisoner-of-war life in Changi were not going to be insurmountable. Not that young Rowland Richards, in twenty-five years of life that included the deafness of his parents and hard grind of the Depression had ever considered any problem insurmountable. Unfortunately, however, the Japanese had a little building operation in mind that was going to take 3,000 men out of Changi to face the worst that the tropical jungle could do to them. Every survival theory that the military machine, medical and social science had evolved was about to be tested to the breaking point.

4

Command Proposes

The temporary accommodation provided for 700 men of No. 2 Battalion, 'A' Force, on the night of their arrival at Tavoy aerodrome on 27 May 1942 was an aircraft hangar. It had a floor of jagged blue metal that had been laid for concreting, but the concrete had not been poured. The British had started the construction of the aerodrome, but the project had been abruptly terminated by the entry of the Japanese into Burma.

The men had route-marched to Tavoy, carrying their packs, twenty-three miles in one day from Simbin, a small harbour on the north-western coast of the Burmese peninsula—the last few miles in a drizzle of tropical rain. Exhausted, wet, hungry, they lay on the blue metal, packed like sardines, listening to the rain and waiting for hot food, but there were no utensils for cooking. They breathed the unpleasant odour of damp, sweaty clothing. Many suffered from dysentery. Most were dull-eyed and in low spirits. Morale was the worst it had been since the fall of Singapore, a reaction to unfamiliar diet, internal upheavals and the shattering of a general confident hope that the move from Changi meant they were on the way home.

Sergeant Jim Lynch, a man of culture and a music lover, shifted on his bed of stones trying to ease his aching body and, out of character for a generally optimistic man, meditated on the mechanics of hate and sweetness of revenge. Devoutly religious though he was, he felt that this was treatment for which vengeance should not be the Lord's; it should be for the Australian fighting men to remember and, some day, to repay.

We Australians are not good haters, he thought. We cannot raise true blood lust. Our attitude towards our enemies is tolerant to the edge of folly. But surely we won't forget the last forty-eight hours, nor the twelve days on the Hell Ships. Every man of us will see that one day Australians are able to repay. In peacetime, let us always be prepared to defend ourselves against aggression. Now we know too well what happens to those who are conquered.

His memory of the route-march was a dazed impression of trudging along with the feeling that red-hot irons were stabbing into the muscles of thighs and calves, and of a pack on burning shoulder muscles that felt ten times heavier than it really was; of rain on his face; of friendly natives who ranged alongside the columns of marching men with gifts of fruit, molasses and water, and charming smiles. Those Burmese, he thought, had discovered the art of cheerful giving.

He caught the eye of his ginger-haired, cheeky faced old school mate, Sergeant Scippy Maher, also 2/15 Field Regiment—a school teacher and keen footballer in civilian life—and managed an answering smile to Scippy's irrepressible grin and wisecrack, 'If we're really going home this is a long, hard way round'. Gunner Blue Lennox, one of Scippy's gun crew, called from somewhere nearby, 'Home was never like this. Think I'll go to India and become a professional fakir after the war. A bed of nails could be quite comfortable.'

Elsewhere in the hangar Private Roy Whitecross wriggled into a pair of dry pants and a shirt, extras which he had carried in his pack, and spread his wet clothes on the stones to make some protection for his body. His youthful hunger had been slightly appeased by the fruit and tea sweetened with molasses that the natives had distributed, and at 2100 hours, when the marchers had halted for evening rice at a Jap barracks outside the town of Tavoy, he had resourcefully volunteered to help get water for tea making as this allowed him to get near the fires and dry off a little. He had been able to get some water (not quite boiling) to steep the tea and sugar in his dixie. But with all the wonderful hopes he had cherished when the rumours about the 'big move' went around Changi now dashed, he felt ready to cry with exasperation.

The Dodger's voice droned a stream of complaint.

'They're inhuman bastards, the Monkeys. We were supposed to be going home and this is what they do to us. And our own brass hats are no better. They promised us we were going home and it was all a lie. They just wanted to keep us from making trouble. You can't trust the blokes at the top.' Dodger always felt better when he had someone to blame.

'Ah shut up,' Whitecross said. 'You were one of the ones shooting off your mouth the most about us going home.'

But he knew that Dodger had not been the only one to spread the furphies back in Changi. Rumours had seethed ever since the first week in May when news leaked out that a large force of Australians was to be moved. They were going out of the country. Around the battalion lines bets were being taken that they would be out of Malaya between 21 June and 12 July. Gunner Beatty, 2/10 Field Regiment, a chief purveyor of furphies, had heard that 8,000 men were wanted for a working party to an overseas destination, thought to be Saigon. It would be cooler and there would be more food. Someone said they thought Lieutenant-Colonel Frederick 'Black Jack' Galleghan, Commanding Officer, 2/30 Battalion, had said a long time before that the men would be home for Anzac Day and it was agreed around the camp fires that Black Jack must have known something. It had been said that there would be an exchange of prisoners, 17,000 Japs for 14,000 Australians. And, Roy thought bitterly, he had believed it. He hadn't even minded that they might go home via Saigon. At least it would be a chance to see a new country. This was letting him do quite a bit of travelling while young, even though the transport and accommodation could hardly be called third class. But now . . . who could know what was really going to happen to them?

Whitecross decided not to think about the twelve-day voyage in the tiny Jap coastal ships, with the ghastly overcrowding, miserable rations, men sick with dysentery. He would concentrate instead on his first experience of the Burmese countryside as seen on the day's march, before the beastly rain started. He had seen several shrines that looked just the same as the Burma Pavilion at Sydney's Royal Easter Show, and priests wearing yellow robes that looked like the robes of disciples in Bible pictures. Tavoy had two-storied houses, shops that were nothing but open ground-floor spaces, and electric lights in the streets. Hardly any of the signs were written in English characters. They can't get many tourists, he thought wearily. He slept. They all slept despite the stones.

Headquarters had known as little about the reason for the move as the men, but the officers had guessed this was not a return home through any exchange of prisoners. The Japanese had advised that men were needed for a work party that would be going to another country by sea. Climate would be better, and rations and conditions would not be so overcrowded. It sounded like a change for the better.

Rowley was too tired to care that rough stones were his bed. The previous night, after disembarkation at Simbin they had slept in a rice go-down on bags of rice infested with weevils. Between the rats and mice that scurried impartially over bodies as well as bags, and weevils that swarmed into hair and ragged clothing, sleep had been almost impossible. At dawn a breakfast of rice and weak tea had been served but he had missed out as he was sorting the very ill from those fit to march that day. Then, during the march he had covered an extra ten miles, he thought, moving up and down the column attending to men who collapsed from exhaustion. He remembered the smiling Burmese who had stopped him and offered him a tiny cup of hot water. Hot water, he had discovered, could be a delicious drink.

Over the last twenty yards to the gates of Tavoy aerodrome, Rowley with his pack, and one Japanese guard with rifle, had been trailing behind the column. Exhausted and stumbling, Rowley imagined that the gates kept receding like the mirage of an oasis. Then he heard shouts from the prisoners who had been halted beyond the gates: 'Come on, Rowley. Don't let the little bastard beat you.' Drawing himself up to his full height, which at least was taller than the guard, he managed to achieve something like a stride and had passed through the gates ahead of his competitor. It had raised a cheer from the men who were almost too tired to move their feet.

He made a pillow of his pack and was almost instantly soundly asleep.

There was a heavy rain shower next morning and all the men who were well enough to move stripped off and revelled in washing in the stinging downpour. Refreshed, Rowley wondered how he could feel so fit after the ordeal since leaving Changi.

He remembered the fifteen days since the 3,000 men of 'A' Force had embarked at Keppel Harbour, Singapore, on the *Toyahasi Maru* of 8,000 tons, and the *Celebes Maru* of 5,000 tons. Two thousand men were crammed into the *Toyahasi*

Maru and the remainder aboard the *Celebes Maru*. They had waited in the hot sun at the harbour while rice and other food stores were being loaded; an opportunity to see the severe damage done to the wharves during the bombardment of Singapore.

Rowley had arranged first aid for a man who had been an object lesson that the Japanese meant to be obeyed, and to maintain strict discipline. Some of the fellows had disregarded the large 'No Smoking' signs and one tough nut was quite blatant about smoking his cigarette in the shade of a wrecked truck until a Japanese pounced and dealt him a sharp slap in the face, physical punishment apparently being the Japanese form of summary discipline. The same guard had returned to patrol the area some ten minutes later and caught another fellow with the lighted butt of a cigarette hidden in his cupped hand. That one was cracked over the head with a strip of wood and knocked senseless. One of the medical orderlies had dabbed the superficial scalp wounds with iodine, and in reply to the man's 'If they do this to a bloke for smoking, what'd they do if a bloke hauled off and punched one of them on the nose?' had warned, 'I wouldn't try it if I were you, mate. These fellows don't play around.'

Remembering that incident, the shoves and cuffs dealt out to men who were slow to carry out orders in loading and off-loading, the bayonet pricks that assisted stragglers on the route-march to Tavoy, Rowley tried to analyse and learn from the experience.

We have to get to know how they tick. These are a non-European race. Their culture is—what—almost feudal? Barbaric. We were so busy studying *Ten Sixty-six and All That* and memorising lists of English kings at school what we were taught nothing about our northern neighbours . . . except that they wore kimonos, cultivated silkworms and lived in a cherry blossom land.

We weren't taught that the picturesque Jap is single-minded, ruthless and places little value on human life.

The first rule for dealing with an enemy is find out his background and how he thinks: then you know his strengths and how to avoid them and his weaknesses and how to exploit them. Maybe you also learn something about your own strengths and weaknesses as far as the enemy is concerned.

One thing is certain, the Japs don't go in for providing any comforts for prisoners of war. We are going to have to live hard.

Three thousand men of 'A' Force had been crowded into two small ships. The holds had been fitted out to carry Japanese troops by the construction of tiers of sleeping platforms with four feet of headroom between each platform. In the tropical heat the holds had been stifling, hot and foul with the odour of packed-in human bodies. The Japanese in charge of the prisoners on the *Toyahasi Maru* had permitted as many men as could find room to sleep on the deck. The concession, Rowley thought, was thanks to Captain Bill Drower, the tall Englishman from Malaya Command who was attached to 'A' Force as Interpreter.

He wore a topee with an English regimental colour patch, his accent was pure Oxford, and his manner was gentle. His Japanese was fluent and he appeared able to argue concessions out of their captors, although in some ways he acted like a typical P.G. Wodehouse 'silly ass'. The appearance and character of the man fascinated Rowley, who thought that anyone as typecast couldn't be for real and would surely be worth studying.

Rowley had set up his RAP on the ship in a small space he appropriated on the for'ard deck. There he sat on his medical pannier and held morning sick parade. Those with upset stomachs were prescribed a day's starvation, those with skin troubles were told to bathe the irritated areas with sea water. His supply of drugs, lotions, powders and pills had to be conserved against real need.

The medical supplies were somewhat more extensive than he had believed when he had packed up his RAP for the move. During the first night aboard the *Toyahasi Maru* his men, starting with Jim Armstrong, Pinky Rhodes, Don Booth and Arthur Baker and continuing with a number of his unofficial TAP medical orderlies, began confiding in whispers to him what they had a 'few supplies' in their kits. They had raided the caches that were buried on the night the POWs had been moved into Changi.

A startled Rowley was aware that the caches were technically the property of Command. 'You mean you bloody well stole stuff,' he almost yelped when the first confidences were made. 'Stole, sir? No way. Those were supplies of yours that you drove in in the truck. Well, we buried them for you, so of course we dug 'em up for you.'

Rowley had to keep a straight face, but when he was able to get a little privacy, leaning over the bow rail at night watching the unfamiliar constellations of the northern stars, he had a quiet chuckle. So maybe his blokes were scroungers, pilferers, sometimes outright thieves, but they were a lot of hard-headed, pragmatic individualists; the sort of Aussie blokes who could fix just about anything with a bent nail and a bit of fencing wire. When they said 'She'll be right, mate', it meant that they could fix things up. They could get by. They could survive. Should he condemn their methods?

Thank heavens for Kyogle and the militia where he had learned to know his countrymen and some of their ways of survival in a hard environment instead of letting himself grow into an ivory-tower academic. He would like—he kicked the rusting rail of the little ship that had once been English and had been sold to Japan as scrap, to be used at last to transport Australian prisoners—he would like to know himself as well as he knew his men.

He thought he was a good doctor, although he hadn't been a brilliant scholar; he thought he had some flair for administration; he thought he had been an efficient soldier. But what sort of man was he? His life had been spent as a student of medicine and gunnery. He should be in private practice now. He was twenty-five years old and a prisoner of war, and God alone knew how long that would last. Perhaps it might only be weeks or a few months as everyone had once thought, but the possibility that it could be years was too dangerous a thought to dwell upon. That way lay despondency and the desire to quit. But quitters were never winners. He wouldn't quit. His mates wouldn't quit. 'A' Force was going to win through.

On 17 May, off Medan in northern Sumatra, three more ships joined the *Toyahasi Maru* and *Celebes Maru*, one of them carrying Dutch POWs who had been taken when Java surrendered to the Japs in March, and a small minelayer was added to the convoy as escort. On 20 May the POW ships anchored off Victoria Point, at the most southerly point of Burma. One thousand of the men of 'A' Force, comprising No. 3 Battalion and sixteen cases of severe dysentery in the care of Captain Claude Anderson, RMO 2/4 MG Battalion, and Captain John Higgin who was an additional physician detailed by Command HQ, were off-loaded, their destination unknown. Already the Japanese were causing the best laid plans to go astray and the carefully planned 'A' Force was being broken up.

'A' Force—the first Japanese work party of Australian POWs to leave Changi—had been designed by the Australian commanders to form a composite group with a suitable cross-section of fighting units so that it would be a fully functional fighting force should it be liberated from the Japanese at some time.

How or when an allied counter-offensive could achieve that liberation could not be guessed at by anyone in Command. Like all the prisoners of war, they were cut off from all real knowledge of the progress of the war in the Pacific and in Europe. The concealed radio receivers, run at the risk of the operators' lives, supplied news that was necessarily vague and perhaps even deliberately misleading, since the enemy tuned in as well. So Command could only hope, and prepare.

'A' Force was under the command of Brigadier Arthur Varley and the battalion commanders were Lieutenant-Colonels George Ramsay and Charles Anderson and Major Charles Green. They had infantry, artillery, engineers, and various medical, Army Service Corps and other units. The battalions became known by the names of their commanders.

Colonel Derham's orders to all medical officers concerned in the move had been distributed on 10 May and were headed 'Move of Detachment of AIF Overseas (Strength 3,000 date 12 May 42)'. Lieutenant-Colonel Thomas Hamilton, Commander, 2/4 Australian Casualty Clearing Station, was to be the Senior Medical Officer. The 2/4 Australian CCS was going, complete with attached personnel, plus twelve ORs drawn from the medical wings of 2/2 Motor Ambulance Company and 2/3 Motor Ambulance Company, two Red Cross representatives, one additional dental officer, and four additional other-ranks dental personnel.

The 2/4 CCS medical staff comprised Majors Hobbs, Fisher, Chalmers and Krantz, with Captains Higgin and White. The dental officers were Captains Simpson and Treleaven, and Mr Don Murchison was to be the senior Red Cross representative. The Field Hygiene Officer was Captain R. Greville and the three RMOs were Captains C.R.B. Richards, 2/15 Field Regiment, C.L. Anderson, 2/4 Machine-Gun Battalion, and G.C. Cumming, 4 Reserve Motor Transport Company. Captain E.M. Lee was Quartermaster. Supplies and equipment to establish a 200-bed field hospital were to be taken.

OR personnel, including a pay sergeant, dental assistants, field hygiene assistants and medical orderlies, numbered 127. With seventeen officers, the grand total of personnel responsible for the medical care of 'A' Force was 144. The chaplains responsible for the men's spiritual comfort were Padres Bashford, Smith and Mathieson.

Rowley, like all members of 'A'Force, had had to say goodbye to good friends who were to stay in Changi. Sid Brien, his lame batman, was left behind, lamenting that he could no longer care for Rowley's leather. The biggest wrench was to part from Des Makepeace, boyhood friend from Kyogle, neighbour at Summer Hill, companion through the militia days and at Holdsworthy, fellow soldier during the Malayan Campaign, and finally fellow prisoner. When and where they would meet again, if they both survived, was in the lap of the gods. It was a breaking-up of his personal group and he felt a little lost although his larger group, the Regiment, was partially intact.

No one, from Colonel Derham down, knew the destination of 'A' Force or the purpose for which the Japanese were moving 3,000 men out of Changi. Speculation was rife among the officers of the medical and fighting units, just as it was among the troops.

When the thousand men, the sixteen severe dysentery cases, and Captains Anderson and Higgin had been detached from 'A' Force at Victoria Point, no information was given as to their destination. They were simply taken off in small boats which disappeared into an inlet in the shoreline. At least the overcrowding was eased a little and the latrine problem (ten latrines on the *Toyahasi Maru* for 1,000 men instead of 2,000), was a little better. It was still terribly inadequate as most of the troops were suffering from some form of bowel disorder due to the diet, overcrowding and the unhygienic conditions of the rat-infested ships.

Although depleted, 'A' Force was still a viable fighting force should the chance come. Yet the departure at Victoria Point caused general unease.

A couple of days later, on 23 May, the Force was further depleted by 1,238 men when Major Charles Green's No. 1 Battalion, plus a half of the CCS, was off-loaded at Mergui, 160 miles north of Victoria Point.

The ship had reached Tavoy anchorage, a further 120 miles north, on 25 May and unloading of the remaining 739 men took place on a night of storm, a nightmare experience of jumping from Jacob's ladders into small boats that were battered against the ships' sides by the high seas, of unloading machinery, and of sleeping in rice go-down.

The worst cases of the 104 sick had to be left at the go-down to await transport; others who could not finish the last couple of miles route-march to the aerodrome had been left at Tavoy gaol.

Rowley realised that establishing a hospital for the sick, and providing adequate facilities for hygiene at an unfinished aerodrome where the main shelter appeared to be the hangar, was going to be a Herculean task. The camp he and Des Makepeace had established at Holdsworthy had been child's play. At least Tavoy would have the remaining CCS staff of sixty-four to set up the hospital. Rowley was hardly into dry clothes before the task began. He was required to join Brigadier Varley in an inspection of buildings allocated for a hospital as soon as he had completed sick parade.

Corporal Pinky Rhodes, with a few medical supplies from the pack that Rowley had carried on his back from Simbin to Tavoy, established an RAP in an engine room near the hangar and there Rowley held sick parade. A doctor without a hospital was like a dog without a tail, he thought, and in civvie life a doctor could at least count on the hospital being a nice, solid establishment of stone or brick which stayed in one place. Army doctors had to cart their hospitals with them, and the 'A' Force hospital, the CCS, was still at Simbin.

'I don't think the Japanese have made ready to receive guests, Captain Richards,' Brigadier Varley said dryly. It was late morning after Rowley's first sick parade at Tavoy and Rowley, Drower and Varley were inspecting the accommodation that the Japanese commander had indicated could be used for the hospital. It consisted of a row of huts of two rooms each with unglazed window openings furnished with atap awning shutters, and with wooden floors. As with all the buildings presently on the drome, it had been built by the British as barracks.

Altogether there was not much to start a POW camp at Tavoy aerodrome: a hangar with rock floor, some wooden huts with atap roofs for troop accommodation, the proposed hospital huts. There were no kitchens, no cooking utensils, no water or latrine facilities, in fact there was no barbed wire around the camp and no guards. Major Itsui, the commander of the Japanese camp a half mile away, simply issued a statement that prisoners attempting to escape would be shot. That, of course, was a contravention of the Geneva Convention, but Major Itsui was a practical individual. How, after all, did one confine prisoners of war if one could not shoot those who tried to escape?

'Tavoy Resort, premier tourist attraction in beautiful Burma. A really desirable piece of real estate!' said Varley, who had been a country real estate and stock and station agent.

'Just like Holdsworthy, sir,' Rowley said cheerfully. 'We'll manage.'

Varley, himself solidly built, gave the slim, young-looking RMO of the 2/15th a sidelong look but refrained from any query.

'Captain Drower tells me Doctors Hayashi and Takuma seem good types, though the medical commander, Dr Chiba, may be a bit hard to get on with. I believe they all speak fair English. Captain Drower will be of assistance if you have language difficulties. I'm sure you can handle the hospital until the CCS arrives in a few days.'

What hospital? Rowley thought, startled. Did a few shells of huts and a bunch of sick men constitute a hospital? There would be no beds, no equipment, no other doctors, no staff and no medicines until the CCS arrived. Only Pinky Rhodes, of his trained medical orderlies, was included in the first group to arrive at Tavoy. He realised that the Brigadier was telling him to establish a hospital single-handed and expected it to be done. Of course it would be done. The RMO was about to start housekeeping again with nothing but the medical pack he had carried on his back and the assistance of Pinky Rhodes.

Rowley found Doctors Hayashi and Takuma actually sympathetic and, in so far as it was possible for them, cooperative. Rowley explained to Takuma the need for proper food and clothing to combat dysentery so that the men could become strong enough to do the work required of them on the drome. The diarrhoea was chiefly dietetic in origin, he stressed, due to an excess of carbohydrate combined with a lack of protein and vegetable. Also, the men were contracting chills from having to sleep in the draughty hangar in damp clothes. Their packs and bedding had not yet arrived.

'The onset of diarrhoea frequently follows a history of being chilled,' Rowley stressed. 'Without proper treatment the diarrhoea develops into dysentery. Men with dysentery are useless as workers.'

Takuma nodded. 'I am aware of the progression. We also have these problems. There are also the problems for us of transport, shortage of medical supplies, food and clothing. I can promise nothing, Captain Richards, but I shall enquire whether there is a remedy for your most immediate problems.'

With that Rowley had to be content.

He organised a detail of four temporary orderlies, led by Duncan MacNaughton, the butcher's delivery boy from Kyogle, to scrounge for cleaning implements (rice straw brooms, cans, anything that could be used to clean the huts). Fit men were set to drying out all spare clothing and blankets so that the sick would have something to lie on. Many of the sick would have to be wrapped naked in blankets whilst their clothes were being dried.

Takuma sent an extra ration of vegetables and a little meat from the Japanese camp for making soup for the hospital.

Rowley numbered the huts, allocated wards for isolation, dysentery, non-dysentery and malaria patients and began registering the sick. He started a hospital register as meticulously kept as if he had still been at the Mater Hospital in Sydney. On the first day, seventy-seven sick were moved into the hospital, including a lad Rowley remembered seeing for the first time on board the *Toyahasi Maru*. He had skin rash then; now he was miserably sick with dysentery. His name was Whitecross and he was only twenty-two. Another severe dysentery case was Chauncey, a fragile-looking man much older than most of the troops. Rowley recorded his age . . . forty-eight. What the hell is he doing in the army at all, Rowley wondered. He should be at home with his slippers on, or working in a factory if he must contribute to the war effort. How was he going to keep Chauncey alive!

Then Major Alan Hobbs arrived from Simbin with three orderlies who were, Rowley saw with glee, Jim Armstrong, Corporal Kemp, the engineer medical orderly, and Gunner Arthur Baker. Old Silver was troubled by diarrhoea, but was ready to be chief nurse despite his discomfort. They brought sick with them, including Major Sid Krantz, 2/4 CCS medical staff, who was a dysentery case. The nuggety Jewish doctor from Adelaide gave Rowley an apologetic grin.

'Sorry Rowley. I'm adding to your work. Dysentery is no respecter of doctors.'

By the time the CCS arrived with their equipment, Rowley had admitted and was treating 195 sick in a functioning hospital. As parties of the CCS began arriving, the energetic Major Alan Hobbs began setting up beds for the men who had been sleeping on the wooden floors. There were sheets and pillows, blankets and most of the equipment necessary to set up a hospital. With relief, Rowley handed over responsibility and returned to his RAP.

Captain Ron Greville, the Hygiene Officer, was attacking the problem of sanitation. Nothing had been prepared by the British proprietors of the drome, or by the Japanese, and for the first three days the improvised latrine was a drain with a Castellani bottle containing disinfectant placed at the entrance to the hangar for washing hands. Created during the Italian war with Abyssinia by Doctor Castellani, the apparatus named after him was simply any bottle with improvised dripper which dispensed a few drops of disinfectant at a time. Since Holdsworthy days it had been as much a part of Rowley's hygiene measures as the daily skin inspections.

Ron Greville obtained or 'liberated' fly mesh, bamboo and atap for the building of permanent latrines—the last word in great Australian dunnies the builders called them—and was supervising the building of the kitchens. The company cooks were performing their own Herculean labours in improvising cooking equipment: drums for boiling water and cooking rice and stews, and mud ovens. By 4 June, when the remainder of the CCS staff had arrived and the main body of the Force moved from the hangar into the billets on the other side of the aerodrome, the entire general area was clean and free from flies. Between holding his sick parades and assisting at the hospital, Rowley found time to inspect the kitchens, the wells that had been cleaned out and equipped with bamboo pipes to provide an ablution area, and the latrines which were placed well away from the living and kitchen areas.

'You're a bloody wonder,' he congratulated Ron Greville and went up to the hospital to express the opinion to Major Hobbs that they should soon see a decline in dysentery and diarrhoea cases.

On 1 June there had been 192 patients, mainly dysentery and diarrhoea, with some URTI and the occasional tinea and fevers. The peak was reached three days later, with 209 patients.

'I thought that in a war we would be dealing mainly with wounds. Plenty of practice in surgical procedures,' Alan Hobbs observed. 'But this man's war seems to be concerned with bowels, skin and throats.'

'What goes down the throats affects what goes on at the other end,' Rowley said. 'We're all going to be dieticians. Authorities on how to provide the maximum nourishment in boiled rice. About the only thing we won't be practising is obstetrics.'

On 5 June the men's kits arrived from Simbin so they had dry clothes and blankets at last.

Rowley was furious. His medical pannier had arrived with the last of the luggage, but it had been looted. He hunted up a clerk-typist at 'A' Force HQ and dictated a letter to the Brigadier with a copy to go to Dr Takuma.

Com 'A' Force

It is with regret that I have to report that the Red Cross Medical Pannier belonging to 2 Bn 'A' Force was looted while in the care of the Japanese soldiers at Tavoy.

The Pannier when unloaded by AIF soldiers at Tavoy wharf was intact and then taken over by Japanese soldiers.

The violation of the Red Cross and the looting of valuable instruments, drugs, medicine, and special convalescent food took place between the time the Pannier was taken by Japanese soldiers at Tavoy wharf and received by me at the main Japanese Guard House at Tavoy aerodrome on Saturday 6 Jun 42.

I hereby lodge the strongest possible protest at the violation of the Red Cross recognised by all civilised countries, and the looting of valuable instruments, drugs, medicines and food which are so urgently needed to treat the sick soldiers in Tavoy.

Herewith approximate list of looted articles.

CRB Richards Capt. AAMC
RMO 2 Bn 'A' Force

Scalpels, scalpel blades, thermometers, eye droppers, medicine glasses, eye baths, an ear syringe of nickel, a nickel steriliser, kidney dishes and a head mirror had been looted, as well as sulphanilamide, aspirin, Dover's powder, acriflavine, Atebrin, sulphanilamide ointment, antiseptic creams, plasters, agyrol, chloroform, salycilic acid, ethyl chloride and methylated spirits. The looted convalescent foods, which were so important for his stomach complaint patients, were coffee and milk, cocoa, chocolate and two dozen bottles of Bonox. Would people at home in Australia, he wondered, who could buy a bottle of beef extract from the grocer's shelf, ever understand the enormity of such a loss?

The hospital was now under the complete control of Lieutenant-Colonel Hamilton and his sixty-four 2/4 CCS, and discharges were winning over admissions. In spite of all endeavours two dysentery cases, Bombardier Osborne and Gunner Austin, had died on 4 and 5 June. Theirs were the first graves of 'A' Force in Tavoy.

Each Company of the No. 2 Battalion who were moved from the hangar now occupied two huts, with their own kitchen and flyproof latrines some 400 yards from the hospital. There was a general shuffling around of troops by the Japanese for which no explanations were given, but the grapevine said that they were going as work parties to Ye, a village eighty miles to the north. Two hundred fit workers followed a party of fifty, under Lieutenant Fleming, 2/15 Field Regiment, and also forty-nine men not yet fit for duties. Rowley was sorry to see Alan Hobbs sent off for Ye.

Doctor Takuma was concerned and sympathetic about the looting of the medical hamper and through Doctor Chiba requested a list of medical supplies on hand and those required for one month. Rowley's list was comprehensive, noting all items that had been stolen.

On Sunday 14 June Doctors Hayashi, Chiba and Takuma visited the camp as a preview for an inspection the following day by Major Itsui. They inspected

Rowley's looted pannier, expressed profund regret that such an incident should have occurred, but were also profoundly regretful that they could make no promises as to how the looting was to be remedied. Following Major Itsui's inspection, however, Dr Chiba again requested a list of medical supplies on hand and those required for one month.

Five days later a few medical supplies, hopelessly inadequate even for immediate requirements, arrived. At least materials were promptly supplied for the making of latrine covers, and following Major Itsui's promise that there would be an increase in the meat and flour ration the Japanese quartermaster visited Brigadier Varley to discuss rations. There was to be an issue of one ox per day to supply the Australians and the Dutch on the aerodrome and in the hospital. One-third of the beast was to go to the hospital. A rice grinder was also promised and arrived on the same day as the medical supplies, but although a slaughterhouse was prepared, no ox appeared for several days.

In preparation for the arrival of the ox, volunteer butchers were requested. The response came from two professionals. Sergeant 'Butch' Hall, 2/10 Field Regiment, was a butcher by trade, and Gunner Kitchener Loughnan, 2/10 Field Regiment, a quiet, good-humoured, easy-going six foot three inches cattleman from Mitchell in Queensland. The youngest of seven children, Kitch's portion of the family lands was a 60,000-acre cattle run which he referred to as 'the little property'. His eldest brother's run was three-quarters of a million acres. Used to cattle on a big scale, Kitch was stricken temporarily speechless when the long-awaited ox was delivered.

'We ought to give that scrubber calf a decent meal instead of eating it,' he said. 'A dingo wouldn't want it.'

It was true that the miserable little Burmese animal was mostly skin and bones and horns, but by the time the two careful butchers had finished dressing it nothing that could possibly be used had been wasted.

'When I think of what I used to throw to the dogs I get a pain in my conscience,' Kitch muttered as he and Butch helped the kitchen staff carry the bones and offal that would make soup stock for the hospital.

The poor beast provided each man with a couple of tiny cubes of meat in the watery stew that was poured over the rice. Yet the prisoners soon learned to welcome the appearance of even the scrawniest ox, since the promise of a beast a day was not kept. Meat on the hoof arrived at irregular intervals.

The cooks were geniuses of invention with the rice. Rice cakes, chapatties, pasties and bread rolls (which were 60 per cent rice) appeared at meals. For the hospital patients exclusively, a small ration of fresh milk was provided in lieu of tinned milk (two issues of eight gallons and one of four), and divided equally between the Australians and the Dutch patients who had been off-loaded from the ships that joined the convoy off Medan.

Although the conditions at the hospital were still in a pre-Crimean War stage, they were slowly improving with some cooperation from the Japanese medical officers. Japanese soldiers were ordered to make more hospital beds according to a model supplied by the CCS, which consisted of rice sacks on frames of bamboo.

The diet was more varied and the numbers of sick were decreasing, but the general condition of the men still showed the effects of the lack of protein and fat. One party of two officers and forty-two ORs who lived close to the Japanese guard house were given handouts of extra meat, fat, vegetables and some spices and curry by friendly guards. That group had the camp's minimum sickness rate: an argument, Rowley said, for the importance of diet and a support of his theory that the best medicine was prevention.

Recurrent malaria was decreasing and Major Fisher advised that all personnel in camp who had previously suffered from malaria (sixty of them), should be placed on suppressive Atebrin—two tablets twice weekly. News came from Tavoy town of a Burmese sick with suspected cholera and Rowley gave a talk to the troops on that killer disease and the hygiene measures for its prevention.

'It's a disease that makes dysentery look like constipation,' he told them.

Several requests were submitted to the Japanese Medical Commander, Dr Higuchi, asking that material be made available for innoculation of all troops. But there was no immediate action. Roy Whitecross thought that cholera must be a terrible disease and resolved to be extra hygiene-conscious. The Dodger remembered that he had been given food in the village and reported for sick parade, convinced that he would develop cholera.

Having sent the Dodger off with a brusque recommendation that he keep away from the village and drink only boiled water, Rowley found himself confronted with little Gunner Chauncey, 2/10 Field Regiment, who was one of the first of his hospitalised patients at Tavoy. Thin and wrinkled, with an unfailingly patient expression, Chauncey was a battler who was always apologetic about turning up on sick parade. He was suffering from diarrhoea again and complained of general weakness. Rowley admitted him to hospital immediately.

Despite the courage of the older men, he felt that he was going to have a hard time keeping some of them alive, especially Chauncey and Gunner Chiddy, 2/10 Field Regiment, who was another wizened battler. Prisoner-of-war life, as the Japs ran it, was not for older men. It was even tough on young men like Gunner Hodgson, 2/10 Field Regiment, who was only twenty-five but suffered severely from arthritis and looked older. Rowley put Hodgson on Light Duties. He was in no condition to shovel earth or wheel skips on the drome. Gunner Dare, also 2/10 Field Regiment and just twenty-four, was sick again with malaria and Rowley hospitalised him. Whitecross complained of blurred vision and painful feet, the symptoms of malnutrition. His young body, used to abundant food, was protesting against the radical change in diet.

Going over the sick parade list in his quarters later, Rowley pondered over the ages of the chronically suffering. There seemed to be two vulnerable groups—the oldest, who were running out of physical stamina, and the youngest who had perhaps not had time to harden. The ages from about twenty-five to thirty-five seemed more tough physically. He wondered if age would be a factor in ultimate survival.

Rowley himself was settling efficiently into POW life in the new environment of Tavoy aerodrome. He felt that he was fairly comfortable in his quarters with a table, stool, valise, improvised bed and sleeping bag; for decoration he had a photo of his young cousin Pat, Aunt Linda's daughter, on the table. The view from his window, rice padi fields stretching to blue hills, and the spectacular sunrises and sunsets of Burma, was nature's own decoration. Gunner Don Booth, 2/15 Field Regiment, who had started as his batman when Sid Brien was left behind, seemed to have developed Brien's tendency to fuss over him.

That the Japanese might vaguely promise and not perform in medical matters was one facet of the captors' behaviour; that they would definitely threaten and did perform in matters of discipline was quite another. The Japanese commander had said that escapees would be shot.

'They're bluffing, of course. Everyone says they can't shoot prisoners of war,' Roy Whitecross said to his mate, Gunner Maurice Barkley, 2/15 Field Regiment, as they shared a rice cake sweetened with chindegar, the coarse raw sugar bought from the Burmese traders. 'A lot of the blokes are saying it wouldn't be too hard to get away and head for the British positions in northern Burma.'

'Don't think about it,' Maurice said. 'I'm not sure they're bluffing at all. Anyhow, a man couldn't take much in the way of supplies. There's nothing to take.'

'Well, they could live off the land. The Burmese are friendly and the Japs can't patrol the whole of the Burmese part of the peninsula.'

'Don't get into bad company.' Maurice was concerned. At twenty-two, Roy might easily be influenced into a disastrous venture.

'No, don't worry. Besides, I wouldn't be much at walking the way my feet play up,' Roy admitted ruefully. 'And my boots are rotten now.'

Sergeant Danaher, 4 Anti-Tank Regiment, and seven of his mates who were not so cautious as Maurice, did plan an escape. The camp seethed, as usual, with rumours based on the small amount of information that was received on the secret radios and embellished with false hopes as it was circulated in whispers from work group to work group, camp fire to camp fire. One furphy was that American troops had already landed in Burma and were only a few days away. It was only a furphy, just as the whispers that everyone would be home by Christmas were without foundation, but Danaher and his friends were not in a mood to wait six months to find out if they would really be free. Their kits had arrived, prisoners came and went without much hindrance among the Burmese villages and the barbed-wire fence on which work parties were labouring had not yet completely enclosed the camp. Perhaps it was the thought of the barbed wire that would soon cut off all their freedom of movement, combined with the exciting idea of joining the 'American liberators' that triggered the escape bid, but their freedom lasted only two days. The Japanese patrols were much more effective than they had thought and the country much more difficult to traverse. Nor were all Burmese friendly.

On the afternoon of the next day, Sergeant Jim Lynch was called out of his work party, where he was cutting bamboo, by one of the Brigadier's staff. He was wanted immediately at HQ. Varley was staring out of his window, grim-faced.

'Sergeant, you've heard that the escapees from the 4 Anti-Tank Regiment have been taken. The Japanese have sentenced them to be shot. I'd like your advice and professional opinion on the legality of such an execution under International Law.'

Lynch recalled the lectures on the Geneva Convention that Captain Curlewis had given in Changi.

'According to the Haig Regulations, sir, it is prohibited to kill or wound an enemy who having laid down arms or having no longer means of defence has surrendered with discretion,' he quoted.

Jim Lynch framed a letter of protest based on International Law and humanity and waited with Varley and Bill Drower for an interview with Captain Chiina of the Japanese Imperial Guards Division. Captain Chiina kept Brigadier Varley waiting—a prisoner of war was lower than a Japanese third-class private, even if he was a brigadier. He finally received Varley for only a few minutes. Chiina was rude and disdainful. 'There is no violation. We follow Imperial Army commands in performing execution. After executions, if you think there is contravention, protest. Execution is agreed to by my superior, Major Itsui.'

A working party was taken out by guards to a spot about a mile from the camp and ordered to dig eight holes, the dimensions of which were suspiciously like graves. That they were graves was certain when Japanese soldiers arrived with stakes which they drove into the ground in front of each hole. With Brigadier Varley, Lieutenant-Colonel Anderson (the escapees' Force commander), Padre Bashford and Bill Drower as witnesses, the eight escapees were executed by firing squad.

A Japanese soldier from Ye suffering with an abscessed tooth brought news of Alan Hobbs—a letter to Colonel Hamilton in which Hobbs wrote:

> There are 250 Australian and 250 Dutch soldiers in Ye. There is no Dutch doctor. There are 15 Dutch and 10 Australians in hospital with dysentery. I can do the work for both if sickness does not increase unduly. The Japanese doctor from the *Celebes Maru* is here and is good to us, but there are few medical supplies here. Could you send as many of the following as possible? This letter is being sent by a Japanese soldier whom I have advised the doctor to send, through you, to Captain Simpson, suffering from a dental abscess.
>
> Ether chloroform }
> Local anaesthetic } I have none
> Mag Sulphate }
> Probe
> Forceps (Spencer Wells—at least 6)
> Scalpel
> Scissors—surgical

Acid Salicylic
Ung Zinci
'Tar Ointment'
Icthyol
Tinct Iodine
Bismuth or any alkaline powder
Higginson's Syringe (if you have more than one) or wound syringe
Aural speculum (one from Bdr Brown's auroscope)

The Japanese doctor will write instructions at the foot of this letter.

Sgd A.F. Hobbs, Maj.

Alan Hobb's letter was countersigned and stamped by the Japanese doctor, Kusunokie, at Ye. It was referred to Dr Takuma by Colonel Hamilton and Takuma made arrangements for the CCS to forward the drugs and instruments. Colonel Hamilton wrote a letter to be sent with the supplies:

> Many thanks for your letter which was received yesterday, by the kindness of the Japanese soldiers. I am glad you are well. I am able to send you practically everything except Chloroform and Tincture of Iodine. I have suggested to the Dutch that they send you an English-speaking medical orderly.
>
> Commander Hayashi and Dr Takuma continue to give us all possible help and the conditions here are much improved. My officers send their kind regards with my own best wishes.

The letter was countersigned by Dr T. Takuma and the medical staff felt a little glow of satisfaction at for once being able to fulfil a colleague's request and even exchange courteous letters.

However, Captain Chiba, medical officer to Japanese HQ, informed Dr Takuma and Colonel Hamilton that Major Hobbs and Dr Kusunokie had plenty of drugs and in any case should requisition through the proper channels. Dr Takuma, on orders, returned the package and letter to Colonel Hamilton.

'Please understand that I regret, most profoundly, Colonel, this unfortunate outcome,' he said gravely.

'Thank you, doctor. We appreciate your considerable effort to be of assistance.'

Hamilton was equally grave, polite and, he hoped, impassive, But he was thinking of Chiba, 'Well, there's sure to be a bastard in the family', and he wondered what Dr Takuma was really thinking.

Bartering with or purchasing from the Burmese villagers—especially one trader, 'Ali Baba', who was permitted inside the camp—added to and increased the variety of the rations. Chindegar, melons and salt could be bought. Suddenly, despite the better food, there was an increase in the oedema usually associated with beriberi. Confronted with puffy facial tissues and swollen legs, the medical officers were alarmed and puzzled. Despite the predominantly rice diet, there should not be an outbreak of beriberi.

The answer was supplied when one of the cooks remarked that it was a waste of energy putting salt in the water to boil the rice when the men then proceeded to sprinkle their ration liberally with salt bought from the Burmese.

'That's it,' Rowley said. 'Salt oedema. They've been without salt since cooking rice in sea water was stopped at Changi. Now they are getting a double ration. More than the kidneys can process.'

Thereafter the camp staff instituted strict control over the purchase of salt. Rowley, who even as a student had found the body's reaction to salt fascinating, recorded the occurrence in his medical diary and made a mental note to enquire carefully into all cases of oedema. Salt oedema, if the tissues of lungs or heart were affected, could be a killer.

Headquarters staff, the medical staff, and even the cooks were all fortunate in that their special skills needed to be exercised in the prisoner-of-war situation just as they had been in a fighting situation: administration, the fight against disease and the provision of food all had to go on. There was a continuity and purpose in their existence. But the troops—their civilian lives disrupted by the declaration of war and then having been trained rigorously to become fighting components of the military machine before being thrown briefly into battle and defeated—were the greatest sufferers psychologically. Morale on arrival at Tavoy had been very low.

In a special medical report covering the three months to mid-August Rowley noted that the low standard of discipline and morale had given place to a degree of contentment which, although there were no cases of severe depression, was purely superficial and would vanish immediately should conditions become worse.

The men developed their own form of defence against boredom and depression. They baited the guards, devised passive resistance to work, talked about food, started recipe books of which the aim was to collect recipes from as many different nationalities as possible (some became fascinating collections of Dutch, Indonesian, British, Burmese, American and even Japanese cooking), began wood carving (chess pieces, some very primitive, some quite intricate, being the favourite subject), traded with the Burmese, read (after the HQ staff liberated books from the remains of the British library in Tavoy and persuaded the Japanese to issue them to the prisoners as toilet paper), and talked about home and girls.

Rowley had always declared that he ate to live not lived to eat, so he was not troubled personally at that stage by the quantity of food supplied, only by the quality as it affected the health of his men. He could not resolve to his own satisfaction the question of whether compiling recipe books was a vicarious pleasure or a form of masochism.

Sex, the favourite topic of conversation in camp at home, was a poor third to food and hating the guards. The cramps of a hungry belly were ever-present, obliterating the libido.

The Japanese command, mindful of the needs of soldiers, had supplied companions for its forces, ladies Rowley referred to as the 'hors de combat', and some of the Korean guards found girlfriends in the Burmese villages. The Dutch who had lived for years in the East Indies and the black Dutch

(Indonesians under Dutch rule), may also have found solace in the villages, but the Australians did not find the idea of socialising with the local belles generally attractive. Apart from the lectures they were given on venereal disease, a bloke could get shot for being caught outside the wire and while that might be a worthwhile risk when trading for food, it wasn't worthwhile for girls. Those who had pin-up pictures passed them around (some making extra food money at an anna a look), but the camp-fire talk, when it concerned women, was sentimental, reminiscences of the girlfriends and wives and mothers waiting at home, not knowing whether their men were alive or dead. The men of 'A' Force hoped that the one batch of POW postcards allowed to be despatched from Tavoy would reach their families.

Missing their own sweethearts, wives and mothers, the Japanese troops were eager to produce snapshots of their families during work breaks. Admiring family photographs, on both sides, became a way of lengthening the rest periods, although one of the guards almost provoked a punch on the nose from Jim Lynch when he declared that his ambition, once Nippon had conquered Australia, was to have sex with the white women. Fortunately for Jim he managed to hang on to his temper, no one having been foolhardy enough to discover by experience what happened to lowly prisoners of war who clouted a warrior of Nippon, even a third-class private. But the talk around the camp fires included some bitter speculation on what Australian civilians could expect if the Japanese were victorious.

A stray American Army flier who had been shot down and had arrived with the group of Dutch prisoners of war from Medan said his command had the situation in hand in the Pacific war zone. Selected madams whose island girls were 'clean' were allowed to carry on their business. 'Rape, like looting, is a kind of psychological release for conquerors,' he said. 'Have you ever seen those drawings of Goya's about Napoleon's war in Spain? "Disasters of War", they're called. Mighty cruel stuff. Our command knows the guys need relaxation. They even have groups of the WACs organised for the job. War effort. Buddy of mine thought it was great until he found his sister was a volunteer. Shows there are two sides to any picture, doesn't it?'

Baiting the guards and dodging work were tricks that went together. One little guard nicknamed George—he thought that was great, 'same name King George'—was a favourite target. His work party reportedly convinced him that his rifle was dirty and conned him into washing it in river water.

The Dodger found it was as easy to lean on a shovel for an hour, just looking as if he was about to work on Tavoy aerodrome, as it had been when he worked for the Department of Main Roads at home. Earth was shifted by the pole-and-bag method. A large piece of sacking was spread on the ground, earth was shovelled onto it, it was tied into a poke by the four corners and then hoisted on to a pole which two men carried on their shoulders. There weren't enough guards to check that every bag contained as much earth as it could, nor to watch the skip line used to cart rock fill for embankments. If a skip accidentally went off the rails, several men were involved for a considerable time in righting it.

Altogether the troops found ways to protect themselves from working to the point of exhaustion and to continue their battle against the Japanese by time-wasting as a way of sabotaging completion of the drome. Even so, runways were eventually completed.

The landing of the first Japanese plane was a complete success from the POW viewpoint. As the aircraft came in the Australian coolies kept up a steady chant of 'Crash you bastard, crash you bastard'. It did.

The ration of oxen eventually became more regular; miserable, emaciated little Burmese cattle that yielded lean, gristly meat with little fat, but increased the protein in the diet. Through the efforts of Don Murchison, the senior Red Cross representative, the Force was permitted to purchase extra supplies (salt, chindegar, pineapples, paw paw, limes, cheese and eggs) from Ali Baba. The money was provided by a fund created by the officers from a proportion of their Japanese work pay—a pittance paid to all working prisoners. Although still of poor quality, the diet was adequate for men working, at that time, six and a half hour shifts of manual labour. The food was monotonous, it was Asiatic, but there were civilians interned in Manila . . . although the POWs isolated from the rest of the war did not know it . . . whose rations were less, and allied soldiers fighting in the steaming jungles of Pacific Islands with nothing to ease the hunger pains of shrunken bellies but small blocks of unpalatable dry rations. 'A' Force was better fed at Tavoy than it had been since the prisoners marched into Changi, or would be ever again while in Japanese hands.

Rowley's special medical report commented on the working parties:

> The effect of five and a half, increased to six and a half hours, work on the aerodrome per day, in spite of the rain and mud, has been most beneficial since the work keeps the men's minds occupied and builds them physically. One sees a body of men who have their ribs and shoulders covered with muscle and fat. Even some of the more puny individuals are now able to cope with the day's work without undue fatigue and men accustomed to heavy work consider they are as fit as they have been since leaving Australia.

The regular hours of the prison camp routine also contributed to the general physical fitness: six and a half hours of labouring; eight to nine hours of sleep; Sunday as a weekly rest day, with time for washing clothes and cleaning quarters. Idle time was taken up with organised concerts, talks and debates.

Although there were four deaths from dysentery and one from possible malaria in 'A' Force, its situation at Tavoy was relatively tolerable. Had the Force been able to remain there or in similar camps for the duration of the war, its fate would not have become part of the history of the disasters and horrors of war. But after mid-August of 1942, Tavoy became paradise lost.

5

Nagatomo Disposes

'Every time we get the housekeeping in order we have to shift,' said Lieutenant-Colonel Hamilton. 'I shifted the Casualty Clearing Station so many times during the Malayan Campaign that I could do it with my eyes closed. The Swiss Golf Club was the most comfortable location . . . not counting shell fire . . . but Tavoy has been almost pleasant. I wonder what the next spot will be like.'

'Another fascinating tourist eye-view of exotic Burma,' Brigadier Varley, in expansive mood, volunteered. 'Charles Anderson is getting his force together as if he was preparing for one of his African safaris. I believe he thrives on discomfort.'

On the plump side when Singapore fell, Varley had thinned a little, the high cheekbones of his round face were more prominent and his lined forehead seemed to have acquired some extra ridges, but he was hardened and fit. His bullet head, on which the little remaining hair was so closely shaved that he resembled a Buddhist monk, was suntanned. Exuberant and optimistic by nature, he was as ready as ever to argue with the entire Japanese Command for the welfare of 'A' Force—or the remains of it.

'We might all get together again. I've heard that Ramsay's force from Mergui and Victoria Point might be sent wherever we're going. Moulmein probably.'

'Then I might have a chance of seeing John Chalmers again,' Hamilton said. 'I wonder if they'll leave Alan Hobbs and "Green" Force stuck at Ye?'

'Poor old Alan, stuck out on a limb at Ye,' Sid Krantz growled around his cigar. 'It was a damned shame he didn't get those medical supplies. I'd like to kick Hyashi in his tenderest spot.'

'We should drink to that,' Varley said. 'I have some saki that cost me a fountain pen. Let me shout a round of drinks for the Tavoy Officers Club, may its atap roof rot and fall in on the Japs. Someone fetch Anderson from his safari planning and find Rowley Richards.'

'That's easy,' Krantz told him. 'Rowley's packing up his RAP with Old Silver Armstrong counting every bottle and scalpel, what's left of them, twice. For a little fella, young Rowley can roar like a wounded bull if the Japs get near his supplies. One thing's as certain as daybreak wherever we go, Rowley will be demanding drains, flyproofed latrines, and boiling water to sterilise the plates and spoons. His methods work, too. Break out the rice wine, Brigadier.'

Some of the missing units of 'A' Force began arriving at Tavoy as from 15 August; first Captains Claude Anderson and John Higgin with a small group of sick and news that about eighty sick altogether could be expected from Mergui. They were followed by two detachments of 200 each, including thirty-seven dysentery patients who had to be admitted to the hospital. One of these, terribly emaciated from continual attacks of renal colic due to calculi (kidney stones composed of mineral salts developed in kidneys and bladder), died shortly after arrival. By the time Major John Chalmers had arrived with the rest of the amoebic dysentery cases from Mergui the Australian section hospital patients had risen from thirty-nine to 111. So the saki celebration for the medical corps had been a little premature.

Eighty men with a history of malaria were put on courses of suppressive Atebrin. By the first week of September, medical supplies were rapidly nearing zero.

Lieutenant-Colonel Charles Anderson, VC, MC, moved out of Tavoy with 700 of his force on 29 September, marching the twenty miles from the aerodrome to the wharf where four barges were waiting to take the prisoners downriver to their ship. The remainder, recently arrived from Mergui, followed later.

Aged in his mid-forties, Anderson was a very precise and correct man who wore round-framed glasses and conscientiously refrained from swearing. Yet he had been a planter in Africa and had led game-hunting safaris before settling in Australia as a grazier. He had won his VC during the Malayan Campaign when he, with a small group of men, had been cut off by a Japanese advance after knocking out ten tanks. After a running battle against surrounding Japs for fifteen miles, they were again cut off and Anderson had continued to fight, refusing to leave his wounded until he was ordered to retreat. He then led the remnant of his men on a break through the Japanese lines.

As RMO, Rowley was with the first group out, in charge of about thirty sick. Apart from them he had to attend to men who became ill on the march from cramps brought on by salt deficiency as sweat poured out of them in the heat. Salt engaged his attention more and more for the human body played tricks both when suffering from too much of it and not enough. Maintaining the right balance was a medically interesting problem on which he was to expend considerable thought.

The remainder of the medical corps and their patients were to be moved in stages, and it was now fairly certain that 'A' Force would be going to Moulmein, Kipling's port of the pagoda on 'The Road to Mandalay'. Where their final destination would be, or why, was in the lap of the Japs.

Jim Lynch, now acting company sergeant-major and company orderly messing sergeant at the princely sum of 10¢ Japanese a day (roughly one and a half pence), marched steadily despite the soreness of a leg he had hurt during a camp football match, carrying a stock of rice, bananas, molasses and sweet potatoes in his sack, trying to occupy his mind with thoughts of his final leave at Holdsworthy only fourteen months before. It had been a perfect Australian winter day with a cloudless blue sky, a bracing nip in the air and the first wattle of the season fluffing out its soft yellow globes. His mother and sisters had arrived with Scippy Maher's wife, Lal, and they'd eaten a picnic lunch and drunk cool beer. Cool beer on a mild winter's day . . . how he would enjoy chilled Australian beer in the heat of Burma! All around them had been the families of the other chaps, some noisy, all happy, he remembered. He had taken snaps and some of them were in his kit now. Now? Nine months a prisoner of war, and where was he headed next?

We were would-be heroes, sons doing their duty, he thought. We've learned to know defeat, hard manual labour and rain—never understood it before—blotting, blinding monsoonal rain that soaks through in a few seconds, stings, and despite the 15 degrees North latitude, freezes. Clothes, hands and feet were always wet, boots had rotted away. The men improvised clogs that were wooden platforms with straps of cloth or scraps of rubber from tyre inner tubes. Some went barefooted but the gravelly mud scalded the soles of the feet and sores had festered between the toes. Rowley Richards' daily skin inspections had helped combat the spread of tinea. Rowley concentrated on detecting skin infections early and carrying out meticulous treatment, but he couldn't beat the Wet season or the lack of footwear.

And rain, incessant rain, upset some of the chaps more than short rations or the slaps the guards dealt out. He'd seen a chap standing in a downpour, drenched, water running out of his hair into wide-open, staring eyes, shouting, 'Stop it, for God's sake. Stop the bloody rain.'

But the only other season in Burma was the Dry—a season of stinging heat and dust. Europeans in the tropics had been certain they could not work in the heat or go out with uncovered heads, a folly sure to lead to sunstroke. But Australians had proved in the founding of their own country, especially in the canefields of Queensland, that white men could work under the tropical midday sun. And now prisoners of war they were proving it again, standing up to the conditions as well as most, if not even better.

I have learned to be a labourer, hardening and strengthening my body, Jim thought. I've learned to live and know what life is, as I never would have done in a secure law practice. I think at last I have my own measure as a man, and I've just turned thirty years of age. I'm full of new hope, new resolutions and I can and will be in the future what previously I only dreamed to be.

Roy Whitecross, a strictly practical survivor and observer, had been rushing around before departure buying sweet potatoes and molasses and gathering the ration of cooked rice and hard biscuits issued by the Japanese. He and one of his mates, 'Hookey', got a pole on which to carry their two kit bags, made heavier by the extra food. Roy had a sore heel and had been working on the drome without boots. Now his feet were soft and his only pair of boots, long unused, were too hard so he was suffering badly from blisters after the first couple of miles. His socks had holes in them, too, and every hole meant another blister.

At Tavoy, Whitecross recorded in his diary, the townspeople were wonderful. 'They rushed out and gave us cigars, mublars [a rice-flour pancake filled with cooked rice, beans, curry, or whatever], sugar cane, cigarettes, bottled drinks and little packets of salted raisins.' Gifts of money, only a few rupees but welcome, for money bought extra rations, were pressed into the hands of some prisoners. All the bigger buildings in Tavoy, he saw, had been gutted by fire from Japanese bombings.

The Dodger developed 'cramps' on the march, in the hope of tossing the doc and getting a lift on one of the supply trucks, but the Dodger had cried wolf too often. When told to march, he reckoned the doc was a real hard-hearted bastard.

A thousand men were transported down-river on four barges, each about forty-five feet long and towed by two launches. The ship that was to transport them up the coast to Moulmein was the *Unkai Maru*. It was about the size of one of Sydney Harbour's Manly ferries and built like a timber cargo vessel with engines and funnel aft and two holds forward. Four hundred men were packed into each of the holds, which were about fifty feet by forty. Each man had five square feet of space for himself and his baggage. The remainder were on deck. When the Japanese wanted to cram a further 400 on board as well, Bill Drower and Rowley protested vigorously and successfully although the Japanese captain, rather at a loss to understand the fuss, pointed out that he had once transported 1,000 Japanese troops on the same boat.

After a twenty-eight hour voyage, the *Unkai Maru* reached the mouth of the Salween River, a muddy waterway about two miles wide, flowing through the monotonous scenery of flat green padi fields with an occasional pagoda or shrine. They marched through Moulmein to a churchyard which was to be their quarters for the night and Roy Whitecross was fascinated by the town's transport system—pony taxis like London hansom cabs and bicycle rickshaws with brightly painted sidecars shaped like small canoes and decorated with flags of coloured cloth and tiny vases of flowers—and by the sight of elephants used for hauling huge teak logs.

Jim Lynch was astonished to find that legal practitioners advertised on bill boards in the streets. One lawyer, Suang, BA, BL, described himself as a 'high grade pleader'. Most unprofessional, Jim thought.

That night the first contingent of 'A' Force camped in the grounds of the Baptist Church, built in 1866. They were able to have their first cooked meal since leaving Tavoy, and slept well. There was also a tap in the church grounds, and Jim, to whom cleanliness was an essential to comfort, was able to have a hasty *tong* (many of the prisoners had adopted the Malay word for ablutions). His shirt had no buttons and no sleeves, and his pants had long ago had their legs hacked off above the knees, but he was clean and felt marvellous. His ego, he decided, was alive and well. The Japs couldn't win as long as men kept their personal pride.

The Japanese treat for them the next day was a rail journey, fifty miles in open cattle trucks with thirty men and their kit bags to each truck. When they arrived at a collection of buildings and what appeared to be the end of the railway line, Jim, dusty and sweaty, said to Scippy Maher, 'Well, here we are. Where are we?'

The place, Thanbyuzayat, was south of Moulmein and the end of the existing railway. A lot of railway stock was stacked around the railhead and there seemed to be considerable activity among the Japanese soldiers and coolie workers. The buildings were long bamboo huts with roofs of atap. Holes, here and there, with atap awnings served as windows and there was a three-foot space between the top of the wall and the roof to allow air to circulate. The air was hot. There was a centre pathway down the length of each hut with raised bamboo sleeping platforms on either side. These were to be the quarters for the prisoners.

'It could be worse,' the American Army flier reckoned. 'I met a guy before I bought it who hadda sleep in burrows in the stinking undergrowth, with spiders and all.' He leered at the Dodger, who had had an encounter with a large spider back at Tavoy and was scared of all crawling and slithering things. Dodger worried constantly about fat, hairy spiders with fangs like curved steel and also the likelihood of an encounter with a cobra. 'The guys in the jungle reckon POWs are on rest leave,' said the Yank.

'Gawd strewth, those blokes ought to cop a few of the bashings we've had from the Nips,' Dodger complained.

'They don't get their heads bashed when they meet Nips. They get them blown off,' Acting Company Sergeant-Major Jim Lynch told the Dodger cheerfully. 'Now what about moving and getting your kit tidy.'

'When I listen to bludging, whinging, points-working chaps like that I feel ashamed,' Jim told the American. 'He'd pinch a sour rice cake from his best mate. I'm glad I've seen real men of the AIF. I hope you don't judge us by men like Dodger.'

'That's okay, buddy,' said the Yank. 'There's some of our guys you wouldn't give two bits for and neither would I.'

For a couple of days the men were able to rest, washing their rags of garments, tidying kit bags, lying on the few square feet of sleeping platform allotted to each man and reading. The books from Tavoy library had been carried one or two in each kit despite the extra weight, and were cherished and carefully circulated. More of 'A' Force arrived, including the lost sheep from Ye with Alan Hobbs in charge of the sick. Rowley, waiting for the CCS to arrive, established a temporary hospital for twenty patients and held sick parade for 188. URTI was on the increase, due largely, Rowley felt, to the large numbers of poor quality Burmese cheroots—gifts from the Burmese in Moulmein—that the men were smoking. Cheroots were made from tobacco leaves put out in the streets to dry but kept moist by passing pariah dogs and the village girls who rolled them on their inner thighs.

He thought it might be an idea to discuss URTI and the possibility of limiting the purchase of cigars with Sid Krantz. Sid was over at the ablutions area, his nuggety body clad only in a G-string and clogs. Water glistened in his curly dark hair and he was shaving the bristle from his olive-skinned face.

As he puffed out his cheeks to take the razor, he rolled his cigar from one side of his mouth to the other.

How stupid can I get, Rowley wondered. When have I ever seen Sid Krantz without a cigar?

'Sid, those damned native cigars are polluting your upper respiratory tract,' he growled.

Sid looked lugubrious. 'They keep the other wogs away. Don't ask me to help limit the cheroot supply my lad.'

'A' Force's first encounter with Lieutenant-Colonel Yashitada Nagatomo was on Sunday, 4 October 1942. The prisoners were assembled for a big parade in the afternoon and the officer who was to hold the power of life and death over them arrived with entourage of staff and interpreter, mounted a platform and proceeded to lay down what Rowley called 'the house rules with all do's and don'ts'. The parade lasted for three hours and it was a measure of the fitness of the prisoners, at that time, that no one standing in the heat of the afternoon sun collapsed. Nagatomo's speech and the interpreter's translation were not audible to all the assembled men, but copies of the speech were obtained through the administration office and circulated during the next few days.

Nagatomo, in dress uniform, wearing white gloves and sword, started by welcoming the men to No. 3 Thai POW Camp and assuring them that the Nippon Government was anxious for their welfare.

'It is a great pleasure for me to see you at this place, as I am appointed Chief of the War Prisoners' Camp in obedience to the Imperial Command issued by His Majesty the Emperor,' he began. 'The great East Asiatic War has broken out due to the rising of the East Asiatic Nations, whose hearts were burned with the desire to live and preserve the nations on account of the intrusion of British and Americans for the past many years.'

'What a load of rubbish,' Roy Whitecross said.

'There is therefore no other reason for Japan to drive out the anti-Axis powers of the arrogant and insolent British and Americans throughout East Asia, in cooperation with China and other East Asiatic Nations to establish the Greater East Asia Co-Prosperity Scheme for the benefit of all human beings, and to establish everlasting peace in the world.'

'Well, I'll be Lieutenant Pinkerton. And he's Madame Butterfly,' murmured the American flier. 'When the robins return in the spring I hope he commits hari kari.'

'During the last few centuries, Nippon has made extreme endeavour to make sacrifices to become the leader of the East Asiatic Nations who were mercilessly and pitifully treated by the outside forces of the Americans and British . . .'

'So that's why Japan attacked China and the Sino-Japanese war has been going on for the last few years. The Chinese must be so grateful,' Rowley observed to Major Don Kerr, the 2 I/C of 'Anderson' Force.

The speech droned on and on: 'You are only a few remaining skeletons after the invasion of East Asia of the past few centuries and are pitiful fact.

It is not your fault, but only your Governments did not wake up from the dreams and discontinue their resistance; all of you will not be released. However, I shall not treat you badly for the sake of humanity as you have no fighting power at all. His Majesty the Emperor has been deeply anxious about all the war prisoners and has ordered to enable opening of our War Prisoners' Camps at almost all the places in the Southward countries . . .'

'Not in Australia, mate,' Jim Armstrong growled.

'The Imperial thoughts are inestimable and the Imperial favours are infinite and as such, you should weep with gratitude at the greatness of them . . .'

Pinky Rhodes wiped imaginary tears from his eyes and blew his nose on the end of his shirt, making a sound that seemed rather like a raspberry of derision. A guard moved menacingly in his direction.

'Be quiet, you clown,' Rowley hissed. 'I don't want to be short one medical orderly with all your training.'

The sun was hot . . . heat had always made Rowley sick when he was a youngster in Kyogle . . . and he gave up concentrating on the speech, catching snatches that indicated Nagatomo was telling them not to complain about lack of supplies as the noble soldiers of Nippon did not have enough of anything for themselves owing to the shortage of materials in distant places. Movement, meeting with friends, possession of money, were all to be strictly limited.

'Living manners, deportment, salutations and attitude shall be strict and according to the rules of the Nippon Army, because it is only possible to manage you all who are merely rabbles . . .' Nagatomo was saying.

Rabble we are and as rabble you should beware of us, Jim Lynch thought. French rabble overthrew a throne and had the heads off a king and a queen. You'll have to grind us into the mud of the jungle, old chap.

Nagatomo was explaining that escape was impossible as there was big jungle to the east, boundless oceans to the west and the Nippon Army in the south, steady and guarding. For anyone who foolishly tried escape, the rules would be severe, as 'ill-omened matters which happened in Singapore shall prove . . .'

He then went on to the great denouement of his speech. The Nippon Army had a great work to promote in the construction of a railway to connect Thailand and Burma. Nagatomo was waxing even more eloquent:

'There are deep jungles where no man comes to clear them by cutting the bases. There are also countless difficulties and sufferings, but you shall have the honour to join in this great work which was never done before, and you should do your best efforts. I shall check and investigate carefully about your non-attendance, so all of you who are really unable to work shall be taken out of labour. At the same time I shall expect all of you to work earnestly and confidently every day.

'In conclusion I say to you "Work Cheerfully, from henceforth". You shall be guided by this motto.'

So we built an aerodrome for them and now we are going to build a bloody railway, Rowley thought. One more prison camp, one more chore, and we are supposed to be honoured. If this joker wants our men to hack into the

Burmese jungle he'd better keep up the medical supplies and the food, whatever his difficulties.

After Nagatomo's speech, orders were circulated that fragmented 'A' Force as the fighting unit proposed by Australian Command in Changi and transformed it into a Japanese work force. The prisoners were reorganised into groups of fifty which the Japanese called kumis with the officer in charge of each called a kumicho. Two kumis constituted a han and the officer in charge was the hancho.

'Green' and 'Ramsay' Forces, originally consisting of 1,000 men each, had been dispersed into other areas and 'Anderson' Force was now also split up from companies into kumis numbered from 37 through to 51. But it still retained a large group identity and even closely knit smaller groups as the kumis were assembled by ranks. All officers were in Kumi 37, warrant officers and sergeants in Kumi 38 and other groups, hygiene, kitchen staff, signals and railway workers in their respective kumis. Rowley realised that the preservation of 'Anderson' Force as a force was supremely important. Without it, morale would have been shattered.

The officers were to receive 25 cents a day pay—equivalent to Japanese pay they were told—non-commissioned officers 20 cents and the other ranks 10 cents. Part of the officers' pay was to be deducted by the patriachal Nippon overlords and banked on their behalf in Japan, and the officers volunteered a portion of their pay to an internal POW Red Cross fund administered by Don Murchison to buy extra food for the canteen, kitchens and hospital. This arrangement left them with the same amount of spending money as the troops, and it could be used to buy chindegar, spices, fruits if any, and cigars.

The first pay was received on 27 November 1942. As a captain, Rowley received Rupees 122.50, of which R52.50 was deducted by the Japs for board and lodging, R50 was loaned to the Red Cross and R20 remained for personal spending.

The remainder of the CCS arrived to take over the hospital at Thanbyuzayat. Brigadier Varley, with a small staff, was to establish his headquarters there and liaise with, or report to, Nagatomo. Sid Krantz was attached to the Force as an additional surgeon and Lieutenant-Colonel Charles Anderson was in command of the Kumis 37 through 51 that would be working on the line through the jungle. Rowley was medical officer of 'Anderson' Force, incorporated now into No. 3 Thai POW Camp.

On 10 October 1942, after almost nine months as a prisoner of war, Rowley had his first experience of the true jungle. The kumis were marched from Thanbyuzayat to the 18-kilometre camp, called Hlepauk.

The scenery changed from flat country with the interminable green of rice padi to ridges covered with thick jungle. Waterfalls glistened silver against the deep green foliage and a clear jungle stream with a rocky bed ran close to the camp—a wonderful place, Jim Lynch though, for the evening tong. Again, the camp was a collection of bamboo and atap huts. The prisoners discovered that native coolies were also quartered at Hlepauk.

Native notions on hygiene, or rather the lack of them, worried Rowley and he went into a huddle with Don Kerr on hygiene measures to protect the Australian force. Don Kerr was dynamic, resourceful and a kindred spirit and together they were to establish standards of hygiene not equalled by other nationalities slaving on the line, although there were to be times when even their best was not good enough to combat the greatest enemy of all.

Although the scenery was magnificent at first sight, the jungle was daunting. Gigantic teak trees grew densely. All the available space between them was a wall of bamboo, varying from an inch to two feet in diameter and anything from ten or fifteen feet to a hundred feet in height. The silence startled Rowley, who had thought of the jungle as being noisy with bird calls and the varied voices of animals. He had also imagined the brilliant colour of many tropical flowers, but nothing relieved the green sameness except the glowing beauty of tree orchids which bloomed in the forest canopy. The jungle was silent and dim like a pillared Gothic cathedral. It was green and noisome with the rotting of vegetation. The tiny voices of mosquitoes shrilled an ominous warning of malaria. The stench from the native quarters warned of dysentery. The humidity warned of fevers.

'Anderson' Force had passed a camp of Dutch prisoners at the 8-kilometre mark and there were rumours that the British were ahead, cutting the track through the jungle and building embankments. The railway was to follow little jungle tracks wide enough to accommodate one or two people walking side by side, or to allow the passage of a small ox cart at most. The jungle foliage high overhead was so dense that it was impossible to see the sky, and little or no sunlight penetrated. Some tracks ran for miles through a night-like darkness.

Part of the railway, Bill Drower said, had been surveyed by the British before the war, but it had been considered prohibitively costly and difficult.

The Japanese were embarking on the almost impossible because they needed a railway that would give them quick access to India from Thailand, avoiding the necessity of the long sea trip down the east coast of Malaya, around Singapore, and back up the west coast of the peninsula. The railway would be some 270 miles long whereas the sea trip was about 2,000 miles. Ten thousand prisoners of war, mainly Dutch and British with a small number of Americans and 3,000 Australians, plus Burmese coolie labour, were to work on the line from the Burmese side. It was rumoured that many thousands more POWs were labouring from the Bangkok side and the two sections of the line were planned to meet just beyond the Burma-Thailand border. 'Near a place called Three Pagodas Pass,' Drower told Rowley and Kerr.

Always quiet, kind, and helpful in a rather bumbling British *Schoolboy Annual* way, Drower nevertheless proved an unusual fund of information. The Australian officers were of the opinion that there was more to Bill than the six foot three that met the eye. It was suspected that they had a British Intelligence agent with them as well as interpreter.

Three weeks after the entry of the 'Anderson' Force kumis into the jungle, another group of Australian prisoners arrived in Moulmein. Lieutenant-Colonel

John Williams' Force numbered 884 men, the majority were of the 2/2 Pioneers from the Middle East, with some survivors from the Australian cruiser HMAS *Perth*, sunk in the Battle of the Sunda Strait on 28 February 1942. There were also 194 Americans from the USS *Houston*, sunk in the same battle. Also with this party was Lieutenant-Colonel Chris Black with 613 men, mainly medical troops who had been evacuated from Singapore before capitulation but had the ill-luck to be captured in Java with Lieutenant-Colonel Norman Eadie's medical unit from the Middle East.

Like Rowley's Malayan Campaign truck, 'Williams' Force were veterans of the fighting in North Africa. They were being returned to Australia, after having survived the worst General Rommel could do, to be remobilised for fighting in the Pacific war theatre. But fortune had not been kind to them. They had arrived at Java only to be caught in the Japanese invasion and made prisoners of war without being able to put up a fight.

Their commanding officer, Lieutenant-Colonel John Williams, an engineer in civilian life, had the lean, clean-cut good looks that would have made a model for an AIF recruiting poster. He wore his slouch hat at a cocky angle, and his wide, clear eyes looked without fear on desert or jungle. He had courage, a determination to fight for the welfare of his men, and an ironic sense of humour. 'Williams' Force, which was to become 'Anderson' Force's companion in suffering during the year ahead, was taken up river by barges to the public wharf, then marched about a mile and a half to Moulmein gaol where Williams was shown some rations and a few tins to be used as cooking utensils.

'Let's get ourselves out of a faint and keep the cooks awake. Between ourselves and the Dutch there are eighteen hundred men to feed,' he told his weary officers. Organisation commenced at 0300 hours and the serving of breakfast lasted from 0930 to 1300 hours with the Japs putting on a head-counting parade in the middle of the meal. The one bright spot of that day was the scarcity of guards. Their guards on the ship had been experts at helping them into the hold with the whack of a rifle butt on the head. 'Which did not improve the temper,' Williams observed.

From Moulmein gaol, 'Williams' Force was moved by rail to Thanbyuzayat, thirty men to a cattle truck, where they were met by Brigadier Varley. Williams and Varley were two of a kind, stubborn Australian soldiers with more guts than diplomatic finesse and a determination to demand decent treatment for their troops. Varley briefed Williams on the railway project and the temperament of their overlord, Lieutenant-Colonel Nagatomo, already nicknamed 'The Butcher', before they were required to present themselves at a conference with the Japanese commander.

The conference lasted five hours.

The Brigadier's son, Lieutenant Jack Varley, met them as they walked back from Nagatomo's headquarters. Although he asked no questions, anxiety was plain on his face. He had already been through a recent crisis between his father and Nagatomo over the signing of a non-escape declaration. Three escapees had been recaptured and condemned to be executed by firing squad and Nagatomo had wanted all prisoners to sign a statement agreeing to the death penalty as a just punishment for escapees. The confrontation between the Japanese and the Australian brigadier had been furious, the Brigadier pounding the conference table with his huge hand.

'That is an absolute violation of the Geneva Convention. I'll not sign an assent to murder, neither will any other prisoner of war under my command,' Varley had rasped. 'Shoot these three men and you'll answer to all the free world when this war is over.'

'We dictate terms,' Nagatomo had shouted back at him in French, 'not you who are the defeated. I order the execution. So it shall be.'

The morning after that scene a sergeant and guards from headquarters had taken Varley from his hut and driven him away in a truck. Jack and the Brigadier's staff had heard only that the Brigadier was to be kept in solitary confinement until he signed the declaration. Major Charles Green had also been confined. The escapees were executed. Four days passed during which Jack Varley could not be sure that his father and Major Green had not been quietly executed also. Then Varley and Green had been released.

Their solitary confinement had been spent in darkness, in windowless cell huts, without bedding and with only a hole dug in one corner of the floor for use as a latrine. They had been allowed one cup of rice a day for food, but no water. Finally, when told that Major Green had become ill, Varley agreed to sign a watered-down version of the original declaration:

'To Commandant General of Prisoners in Thai: The underneath signed shall hereby solemnly declare that they shall act in conformity with every regulation and obey this declaration and further promise not attempting escape nor to help others in such project.'

The signing allowed Nagatomo to save face while Varley was sure that duress could be proved and the declaration would be worthless before any future war crimes tribunal.

Immediately on top of that crisis had come the arrival of 'Williams' and 'Black' Forces and Nagatomo's wish to separate the officers and their men.

'It's all right, Jack,' Varley told his son. 'We had some tense moments but we won this round. The officers are to live with their men. If Nagatomo had had his way the morale of the fellows would have been shot to pieces. We'd have had no way of protecting them. His Mightiness is furious now,' he added thoughtfully, 'but he'll calm down and I think he'll have more respect for us. If I'm getting the measure of Japanese thinking they despise people who take a soft line and keep backing down, but they respect strength in an enemy. It's something to keep in mind.'

'Williams' and 'Black' Forces endured the same parade and 'welcome' speech by Nagatomo that had been imposed on 'Anderson' Force, and were then reorganised into kumis. Armbands were issued to the officers, orderly room staff, hospital staff and the cooks. The men detailed for the cutting of firewood, drawing water and building huts, the Buppin Kakari, were given armbands also and the other prisoners were issued with prisoner-of-war numbers. At the beginning of their military careers they had been given numbers and drilled to answer to them as automatically as to their names. Now they had Japanese

POW numbers to memorise as well. Numbering was to be in Japanese and all measurements in metric.

The survivors of HMAS *Perth* and USS *Houston*, were also numbered and assigned to kumis. The *Houston* survivors were sent off to a kumi of their own countrymen and the *Perth* men were assigned to 'Williams' Force. Black's group had the majority of the 6 Division Motor Transport Company, the Engineers and the RAAF prisoners. The sailors thought it was a queer turn of fate that had brought them to coolie work in the Burmese jungle and *Perth* Chief Petty Officer, Vic Duncan, remarked to his mate Chief Petty Officer Tommy Johnson, 'I thought I joined the Navy to see the sea and girls in every port, not bloody teak trees and bamboo'.

'Williams' Force was transported by lorries to its first work camp, Tan Yin, at the 35-kilo mark from Thanbyuzayat, jolting over the jungle tracks at 25 mph in a cloud of dust and flies. They arrived at Tan Yin by mid-morning and were handed over to the new masters of their fate, a group of surly-looking guards, all Korean privates third-class whose idea of maintaining discipline and giving commands was to shout at the tops of their voices and clout, with hand or rifle butt, any prisoner who moved too slowly for their liking. Through shouts and gestures the commander of the guards conveyed to John Williams that his men were to start work immediately clearing a parade ground, and it was to be finished by 1600 hours . . . five hours with no time out for a meal!

'We're supposed to clear that by mid afternoon?' Williams stared at the thickest timber he had ever seen, then at the tools the guard commander was indicating—crosscut saws, axes, picks, shovels and machetes. Even with nearly 900 fairly fit men it was a big order. He remembered his first sight of Australian prisoners working like coolies as the trucks had passed by the 18-kilo camp at a place called Hlepauk: sunburned men in ragged shorts or no more than loin cloths had been swinging picks. shovelling earth, and carrying the dirt in bags slung from poles each of which two men toted on their shoulders. Obviously the only tree felling and earth-moving machinery in this jungle was flesh and blood.

'Not possible by 1600 hours,' he told the guard.

'*Shigoto* now. All *shigoto, speedo*,' the guard grunted. He walked away. At least he had not insisted on a particular finishing time.

The area was levelled sufficiently to hold a parade by 1830 hours and the head-counting and shouting out of rules of conduct went on until after dark.

Night came quickly in the jungle which was gloomy in daytime anyway. The cooks had had the afternoon to get organised at least and the Buppin Kakari, or camp workers, had gathered firewood. Cooking utensils supplied were two buckets per fifty men and fourteen cast-iron frying pans for cooking the rice ration. There were no containers for boiling the water which had to be drawn from a nearby creek. So the evening meal was rice and a watery stew . . . the meat of one cow, weight about 300 pounds, being the daily issue, they were informed, for the 900 men of the entire force. Four or five ounces of meat with a few chunks of cooked cucumber barely flavoured the half dixie of rice issued to each man. The men ate huddled around camp fires which offered the only light. And the darkness beyond the circles of firelight seemed even a deeper black by contrast. Their new prison camp was hardly home from home.

Perhaps tomorrow they would have a chance to settle into huts, where 200 were expected to exist in accommodation designed for half that number, and things would be better. But that first night they were too exhausted to desire anything more than their few feet of space to lie down and sleep.

At Hlepauk the 'Anderson' Force kumis had been rock chopping and carting, building a level crossing.

Jim Lynch, whose eye for beauty was undimmed by the work in heat and dust, enjoyed the contrasting colours of green jungle and, westward, back towards Thanbyuzayat, the brown of padi fields where the ripened rice was ready for harvesting. The natives were reaping the rice and laying it in bundles. The only conflict that day had been when a guard (about five feet tall) decided to slap a driver who happened to be six foot five: the driver had to bend down so the Nip could reach to slap his face. Jim enjoyed his evening tong in the river and thought that he was fairly fit despite an attack of fever one night which had been staved off by a prompt dose of quinine administered by Old Silver at the RAP.

Rowley had set up his RAP in half of a smaller hut. With a camp strength of 600 at Hlepauk, he had thirty patients hospitalised and sick parade figures had climbed to 170, then 220, or about 35 per cent of the force. There were thirty cases of malaria and his supply of quinine was low. By the light of half a candle he wrote up his medical diary. Medical supplies provided by the Japanese for one month were: 24 x 2-inch bandages, 6 lint, 1 oz. potassium permanganate, 2 packets of aspirin, 4 of magnesium sulphate, and 2 tins of throat pastilles.

The approved ration of rice 700 gm, meat 100 gm, vegetables 500 gm and sugar 10 gm was not anywhere near sufficient for men doing hard labour and the sick parades showed an increase in the early signs of vitamin B complex deficiency; neuritis, the inflammation of nerve endings that caused pain in the legs and feet; amblyopia, a dimming of vision although there was no discernable change in the eye itself; central scotoma, blindness of central vision; and the swollen tongue and burning sores in the corners of the mouth that were signs of pellagra, together with patches of inflamed skin over the body, especially around the testicles.

Historically, pellagra in European armies seldom progressed beyond the dermatitis and mouth sores stage, but in this tropical climate, with the work load increased and malnutrition evident, it could become a disease that maimed and even killed; killed painfully and without dignity by robbing the victim of sanity. First symptoms might appear to clear up, then the disease could return months or even a year later if the cause was not rectified, causing blindness, dementia as it attacked the brain cells, and death. Dermatitis . . . Dementia . . . Death.

Everyone, including the Japs, is afraid of cholera and of beriberi, Rowley thought, but they haven't recognised pellagra as a danger . . . not as immediately vicious and spectacular as cholera, but a killer. Somehow we have to get it through to Nagatomo that there won't be enough fit men to finish this damned railway for his engineers unless we have medical supplies and better food. But how do you get through to the Butcher at all? Or to any of the guards?

Signaller Larry Oakeshott, 8 Division Signals, had developed appendicitis only three weeks ago but Corporal Tanaka had refused Sid Krantz's request that he be sent immediately to Thanbyuzayat. So the operation had been performed on a makeshift bamboo table in the hospital hut, with Krantz's surgeon's hands moving fast and Rowley administering the anaesthetic of which they had only a thirty-minute supply. The operation was finished with only three minutes supply of anaesthetic remaining. It was a gangrenous retrocaecal ruptured appendix. There was nothing more Sid Krantz as the surgeon could do, and there was little Rowley could do with his limited supply of drugs except administer M & B 693 and morphia and encourage the lad to battle on against peritonitis. Jim Armstrong and Pinky Rhodes were devoted nurses, but on the twenty-first day after the operation, Larry died of peritonitis and was buried that morning.

Rowley put down his pen and stared at the candle flame, already guttering. Now that 'A' Force as a fighting force in the military sense had been fragmented, distributed in groups through miles of jungle, there was only one fight left—to survive until the war or the damned railway was finished. Command back in Changi had proposed, but Colonel Nagatomo had disposed.

But we're not without purpose, we're not beaten, Rowley thought. We can live in spite of the Japs and this bloody jungle. Just live, day by day by day.

At Thanbyuzayat Colonel Nagatomo was giving a bridge party. He had decided to learn how to play contract bridge and to play well, and by doing so demonstrate the superiority of the Japanese over Western intellects. He spoke fluent French and was learning English for the same reason, to demonstrate racial superiority. The British people who could speak only the English language and made ludicrous efforts to pronounce a few Japanese words lost face before his superior ability. So he had commanded the 'recreation' officer to give him bridge lessons. He studied the rules of the game with the tenacity of purpose that his race applied to anything they undertook and that had brought them out of the Middle Ages into the world of Western technology in less than a century. When he made a mistake in the game he considered it carefully and he rarely made the same mistake twice.

The men he had commanded to be his guests for the bridge evening were Australians Captain Claude Healey, the 'recreation' officer, Captain Ray Griffin and Brigadier Varley. The courtesy with which Nagatomo welcomed them to his quarters in the administration building gave no hint that only recently he had kept Varley and Major Green in solitary confinement on starvation rations. The table was set up with two decks of cards and a bowl of some form of Japanese sweetmeat. Before the game commenced one of his staff offered them tiny bowls of pale Japanese tea. A single spray of the small, delicately coloured tree orchids had been arranged on a side table in the elegant, traditional Japanese manner.

Varley was a poker-faced card player but he found the Butcher was poker-faced too, bland, careful and with a fantastic memory for the cards. Although Varley's eyes remained on the cards and registered no variation of his thoughts, his mind was working overtime and his thinking ranged further than the bridge game.

The bastard is good, he thought, too good. Behind all the temperament, the strutting and the bombast there's a keen mind and a determination not to be beaten. We were told they were barbarians, a backward race, practically monkeys just down from the trees trying unsuccessfully to imitate their Western betters, but it's time we stopped believing our own propaganda and started finding out what they are really like. When they stop imitating and start innovating instead, God help the Western world and God help Australia if they ever get a foothold on our shores. With the Americans and British we can win this war yet, but we'd better not imagine that the Japs will lie down after a beating. Those of us who survive are going to have to remember everything we learn about them.

Varley discovered that he had not been concentrating on the game in hand.

'My trick, I believe, Brigadier,' said Colonel Nagatomo, permitting himself a very slight smile.

6

The Gates of Hell

Apart from a murder, Christmas 1942 gave no indication that Hlepauk was the gate of Hell through which 10,000 Australian, British, Dutch and small groups of other nationalities were to enter into an eternity of suffering.

During November the cold of the nights had increased and the Australians adopted the procedure of putting on all their clothes to go to bed, then stripping off in the morning to shorts or G-strings and hats to go to work. The quality and quantity of the rations provided by the Japanese deteriorated but the prisoners devised ways of augmenting the food. The canteen fund to which each man contributed a proportion of his pay was managed by Don Murchison, the senior Red Cross representative attached to 'A' Force, who worked with Don Kerr, making purchases of sweet potatoes, eggplant, cucumbers, eggs, limes and chindegar from the Burmese wayside traders.

Rowley and Sid Krantz contrived to find a continual supply of men needing dental treatment by Captain Simpson at the base hospital at Thanbyuzayat, a ruse which allowed them to purchase and transport supplies from the fairly well-stocked canteen at base. The men also made private purchases from the Burmese jungle villages, taking the risk of being caught outside the environs of the camp and condemned as would-be escapees in order to trade.

The Indonesians from the Dutch colonies shared their knowledge of edible plants in the jungle and parties were allowed to go out to gather a type of wild spinach which was not particularly palatable used raw as a salad and became a less palatable, slimy green mush when cooked, but it was at least a much needed green vegetable. The berry of a shrub which the prisoners called 'wattle berry' and was said by the Dutch Indonesians to be useful in the treatment of pellagra and beriberi, was a source of vitamin B and was gathered for the making of jam, with melons and chindegar.

The meat ration, of scrawny Burmese cattle, was supplied on the hoof, and as there were no yards to pen the animals the meat had a tendency to escape. The older and least fit men on Light Duties were detailed as cattle watchers, a job considered cushy except for the abuse from their mates and cuffs from the guards that rained on the cowboys whenever precious animals took to the jungle. The guards viewed the escape of cattle as entirely the fault of the prisoners and escaped meat was not replaced.

Assessing the temper of the new guards and learning how to deal with them was a necessity. These were all Koreans, third-class privates of generally low mentality, scorned by the Japanese as little better than coolies although they wore the uniform of the Nippon army, and brutal in their treatment of the prisoners . . . some from fear of their Japanese superiors and some because of a natural cruelty. There were very few genial types like little George at Tavoy, but there were two extremes: those who, in confiding mood, claimed they were as much the slaves of the overlords as were the prisoners of war, and those who delighted in sadistic treatment of the white men at their mercy.

Each of the officers of the Australian forces developed his method of dealing with the Nips according to his character and estimate of the situation. The toppling domino effect in the Japanese chain of command was obvious. The engineer officers, of whom the prisoners saw little, were committed to building a railway through some of the worst jungle in the world: in the fulfilment of that immense undertaking flesh and blood was expendable, and excuses for failure were never acceptable. Corporal punishment for failure was severe and immediate, with beatings going down the line from senior officers to the Korean guards whose work parties had failed to fulfil the day's quota. Right at the bottom of this feudal structure were the despised prisoner-serfs.

Lieutenant-Colonel Anderson believed in diplomacy with a smile, negotiating with the most senior Japanese available so that he could understand their requirements and, where possible, placate the uncertain Nipponese temperament in the hope of avoiding explosions of wrath that would create damaging repercussions for the whole of 'Anderson' Force.

Lieutenant-Colonel Williams was a straightforward soldier, without guile, passionately devoted to the welfare of his men and prepared to argue every point on their behalf even at the cost of his own physical suffering—an attitude which bought him trouble and personal beatings that he could have avoided.

Rowley, as the RMO attached to 'Anderson' Force, was responsible for the health of about 700 men, which meant, he was beginning to realise, their very lives. They were not only his comrades in arms to whom he was loyal as a soldier, they were his patients, every single one of them. He saw them not as a force, but as individual organisms of flesh and blood, brain and complex nervous sytems exposed to attack not only by their overlords but by extremes of heat and cold, rain, malaria, bacteria and viruses, malnutrition and avitaminosis. He was a medical officer trying to be a doctor with precious little equipment other than his wits.

Although he gave the circumstances of his life in connection with the present situation little conscious thought, they had prepared him for it: the courage and perseverance of his handicapped parents, his necessary self-reliance, the rough and tumble of country life and friends at Kyogle, the wisdom and guidance of Arthur and Linda Lewis, the decision to make his way through the ranks of the Survey Regiment instead of joining the University Regiment. He could work points, scrounge and improvise as well as any of the men. By studying the deviousness of the captors he began to learn how to employ deviousness and guile to match the little devils at their own game. He had a henchman in Don Kerr, the 2 I/C who kept a few tricks and stratagems up his tattered sleeve. Jim Armstrong, in his quiet way, also knew how to play the ball game and never fumbled a pass from the doc.

Thanbyuzayat was another world where Varley matched wits with Nagatomo, and Hamilton and his staff had an established hospital to deal with the sick sent back from the work camps and at least some cooperation from their Japanese opposite numbers.

Corporal Tanaka, commander of the guards, was a not unpleasant person who attempted to do his duty conscientiously and was disposed to be reasonable when not under pressure from the engineers. Unhappy with the miseries the jungle imposed on captives and captors alike, he missed his family and was probably a kindly husband and father at home. A Korean, nicknamed the Jeep, was endowed with a large pendulous nose on a flat face and a large pendulous belly on a short thick body, which gave him a ludicrous appearance and made him a natural target for the pranks of the Australian troops. When pressed too far, however, he revealed that he was also endowed with an explosive temper.

The Black Prince, nicknamed for his very dark skin, and the Storm Trooper were naturally vicious and sadistic. A giant among his fellows, the Storm Trooper was over six feet tall, well-muscled with the square jawed, hard-boned face of a pugilist, a low brow and an upstanding brush of thick black hair. While the prisoners disliked their guards as a matter of principle, they learned to hate the Storm Trooper. 'Peanut' Tomoto seemed harmless enough and was, at times, ingratiatingly friendly but with tendencies that made the prisoners wary in his vicinity.

Christmas Eve started badly with an assault in the morning on Captain Bill Drower. As rations became more meagre and the black market among Nips, natives and some sharp dealing prisoners became dearer, anyone with small valuables hidden in his kit began selling them for extra food money. Bill Drower had a prized possession, a memento of that far-off, cultured and elegant life in pre-war England before Hitler's *Luftwaffe* tore the fabric of British civilisation at the seams. It was a silver cigarette case, for which value could be obtained by selling it to the Japanese. Bill had finally resolved to part with it but, disdaining to take part in the trade with the Japanese, had left it at base with one of the other national groups for disposal. It had been bought by a Japanese officer who was delighted with his new toy until, by fiddling, he broke the clasp. The Japanese then claimed the case had been faulty to begin with, set about discovering the identity of the owner and demanded his money back. Bill refused, insisting that the sale had been a fair one.

There, he had thought, the matter would end, but it exploded into violence when he was summoned to the guard house at Hlepauk and jumped by ten Nip guards. Although hitting back when struck by a soldier of Nippon as a disciplinary measure had never been tried, since the consequences would probably be disastrous, a ten-to-one attack over a personal dispute was too much for Bill's well-schooled English restraint. He fought back and was dragged down fighting. When his attackers put the boot in he knew he had to yell for help or risk being kicked to death.

The voice of Bill Drower, kindly, unobtrusively helpful friend, was recognised and the news was shouted through the camp. The atmosphere was electric, the Force, on morning parade, was ready to move in a body against the Nips at the risk of being gunned down. Charles Anderson sprinted for the guard house while Majors John Shaw and Don Kerr quietly moved to control possible mutiny. There was no smiling diplomacy about Colonel Anderson, VC, as he got between his interpreter and the Nips.

Bill Drower had sore ribs, bruises and a few abrasions which Rowley treated with antiseptic. The English interpreter's slighly shaken calm he treated with a tot of brandy from his precious small flask. Then the men of 'Anderson' Force not required for work on the line settled down to their preparations for Christmas Day.

With over 200 of a camp strength of 600 men on sick parade, mainly tropical ulcer cases which were early signs of vitamin deficiency, some dengue fever and about thirty malaria, Rowley marvelled at the effort and energy mustered to organise the Christmas celebrations. Even old Chauncey, although hospitalised, was decorating wrapping paper. Some of the Light Duties men were dressing a Christmas tree, using stubs of candles, scraps of foil saved from empty cigarette packets and decorations cut from paper acquired from the administration office. There were to be Christmas presents for every man, surprise parcels of which the diminutive Jackie McKone, who was recovering from dysentery and working in the RAP, seemed to be the main organiser.

An ex-jockey and batman to Lieutenant John Ross, McKone was the sleight-of-hand expert of the 2/15 Field Artillery Regiment. At camp in Holdsworthy he had refrained from taking part in the two-up games in barracks but would lie on his bunk, seeming to be reading or dozing while banknotes went into a pile in the 'pot'. The gamblers gathered in a circle around the 'spinner', the heads went up as the two pennies were tossed into the air, the heads went down again to see how the coins had fallen, the lucky ones scrambled to claim their winnings and found the pot was empty. Then McKone produced the money from under his pillow, filched from the pot while all eyes were on the spinning coins.

The what and how of the surprise parcels was Jackie McKone's secret until Christmas Day.

Rowley's Christmas Day, 1942, began just after midnight. He was awakened to attend to a Nip guard who had abdominal pain and thought he was suffering from appendicitis but was suffering from a surfeit of unmasticated bananas. Having prescribed a 'No. 9 Tablet' (reserved for constipation of extreme obstinacy), Rowley found that he did not want to sleep. To go back to his bunk would mean staring into the darkness, wondering how his loved ones at home were managing. As far as he knew no lists of prisoners of war taken in the Malayan Campaign had been released in Australia. His parents would not know if he was alive or not. Nor did he know if they or the Lewises were well. It would be terrible to get back to Australia and find that someone had died.

There were others awake in the camp on this cold, dark Christmas pre-dawn when no stars were to be seen because of the canopy of jungle. The light of an oil lamp flickered in the hospital hut where Jim Armstrong was watching the sick. Signaller Beckford, a malaria patient, tossed and muttered, breathing heavily in restless sleep, worn and emaciated after repeated bouts of the disease. Corporal Rooney, over the crisis, was sleeping peacefully. Jim was sitting on the edge of his cot winding old bandages that had been boiled for reuse. He was being helped by a visitor, Corporal Bob Kemp, a medical orderly from the 2/12 Engineers.

'Everything all right, Jim?' Rowley asked quietly.

'Quiet now, sir. Beckford's pretty low. He doesn't seem to want to fight anymore. Everyone else is fair enough.'

'Would you two fellows like a brandy?'

Bob Kemp stared. 'You're not joking, sir?'

Rowley produced his flask. 'Behold, I bring tidings of comfort and joy. One flask, veteran of the war in Malaya, refilled before we left Singapore and conserved to toast our homecoming at Christmastide. But not this Christmas unless our boys and the Yanks get on with the big counter-offensive within the next few hours. Let's drink to being home for Anzac Day. We'll march with the veterans.'

'We'll drink to that.'

The early Christmas drink was just a sip each from the flask, then Rowley left Armstrong like a shepherd watching his flock and went to visit the kitchens where the cooks were working all night.

'Did the pigs Nagatomo promised for our Christmas pork turn up?' he wanted to know.

'Not bloody likely. But not to worry, the Hlepauk Ritz will turn on a Christmas dinner you'll never forget. Would you mind shoving off, sir. The menu is supposed to be a surprise.'

'Not until I've tasted those things that look like drop cakes . . . with jam even. In my capacity as your Regimental Medical Officer, I should test the quality of the food,' Rowley said with spurious dignity.

He was given a mouthful of fried rice cake with a smear of wattle berry jam and decided that, finally, he would be able to fall asleep without thinking.

Jim Lynch lay awake. He was feverish, probably a touch of malaria again, but hardly enough for him to report sick. The rest over Christmas day would fix it. He was a bit nervy and the old temper wasn't the best these days, but most of the fellows' tempers were fraying with the hard work and the cold and the rations that were never filling enough. He would think about warmth and beauty and the eternal symbol of hope . . . the Lamb of God who taketh away the sins of the world was born on Christmas Day. Perhaps mother and the girls were attending Midnight Mass in St Mary's Cathedral in Sydney. It would be warm, and glittering with many candles, light flashing on the gold embroidery of vestments and on the altar vessels, and the voices of the choir would rise into the dimness of the Gothic vaulting.

Roy Whitecross was awake because the ulcer on his leg was throbbing and he had a toothache. He was thinking how strange it was that he had been sure, when at Tavoy, that they would be home for Christmas, yet he didn't feel bitter that this Christmas was being spent in captivity after all. He was coming to accept captivity.

The Dodger was awake and blessing his good luck. He had sold a dozen cigarette lighter flints to a Burmese for three eggs, only the flints were three genuine ones and nine tiny bits of wire. He would use the eggs to make puddings of ground rice and chindegar which he could sell for 5¢ each and the profit could be ploughed into native cigars. Further, he had acquired a shirt that someone had been careless enough to discard down by the river. It was faded, buttonless and sleeveless, like most shirts in camp, and hardly identifiable in case anyone had the gall to try to claim it. He didn't think anyone would.

Corporal Tanaka was also awake. He was probably thinking about his family whose photos he had proudly shown to the prisoners. Despite the encouraging news of continual victories for the glorious armies of Nippon, he suspected that the American submarines were affecting Japanese shipping in the China Sea and food might not be as plentiful in the homeland as one could desire. However, his quiet, competent wife would contrive to take care of the family.

Reveille on Christmas morning was at 0800 hours. Padre Bashford conducted the Protestant service and Padre Smith—known affectionately as 'Pessimistic Pete' because, despite his great faith and hope in matters spiritual, he was convinced that something would always go wrong in matters temporal—celebrated the Roman Catholic Mass.

With breakfast the kitchen staff distributed the menu for the day which had been printed on small individual cards:

Breakfast: Nutty Delight (In fact the usual porridge of burnt ground rice know as Burgoo). Burma Chantilly Sauce (a concoction of chindegar and water).

Dinner: Roast Sirloin of Zebu with Juice and Vegetables; Baked Pomme de Terre, Baked Pumpkin, Haricot Beans and Steamed French Radish.

Sweets: Queen Cakes (a dough of rice flour and yeast dropped in lumps into boiling water) served with melon conserve.

Tea: Braised Haricot Steak, Chipped Pomme de Terre, followed by Steamed Rice with Melon Conserve; Black Coffee (made from ground baked rice partly burnt to give the 'coffee' colour, if not flavour).

None of the titbits for the Christmas meal had been provided by the Japanese but had been purchased from Burmese wayside traders using reserves from the canteen fund.

After dinner Rowley, Sid Krantz and Don Murchison relaxed with a real cigarette—a Players—of which Sid and Don shared a small tin. They then went for a stroll around the camp and listened in to a quiz contest between two of the kumis. Jim Lynch had forgotten about his mild attack of fever in his delight that his team, led by his mate Scippy Maher, had won again: it was never defeated by any other kumi. Quiz contests, with betting in native cigars, were great morale boosters and dictionaries and small encyclopaedias from the Tavoy library stock were the most sought after books in the camp.

The 22 Brigade Band, mostly members of 'Ramsay' Force, were brought over from Tan Yin, where they had been taken the day before by Lieutenant Yamada to give a Christmas Eve performance for 'Williams' Force. They repeated their concert, the inevitable 'Waltzing Matilda', Christmas carols, and marches including the American 'Wings Over the Navy' and the English 'Colonel Bogey'. They finished, although Hogmanay was six days away, with 'Auld Lang Syne'.

Rowley—remembering the post-prandial conferences of Uncle Arthur and E.N. Larkin, father of Norman, in Kyogle after Christmas dinner—took a nap.

After tea the amateur performers in the camp gave a concert, the candles of the Christmas tree were lighted and 'Slapdash' Purdy played Father Christmas, handing out the presents, unwieldy parcels containing a cigar, a banana or some small item of the recipient's own: cigarette case, socks (reserved for wearing out of prison camp), pen, or a prized trinket which had been secreted in the innermost recesses of their kit bags but had not escaped the expert explorations of Jackie McKone.

Colonel Anderson was given a beautifully hand-decorated menu with a sketch portrait of himself and a space for autographs, on behalf of the men of Anderson Force.

A driver from Tan Yin brought news of 'Williams' Force. They had at least received the three pigs promised by Nagatomo, and also a small orange each, but the pork had been boiled in drums as there wasn't enough metal to be had for making egg slicers let alone oven trays for baking. Apart from the band concert there had been no organisation of entertainment by the Japanese.

We've done well, Rowley thought. Thank heavens morale is keeping up. Whatever that unknown quantity is, the '*s*' factor that survival is all about, maybe we have it, maybe I'll know before this is all over.

Boxing Day was another work day. There were 150 on sick parade but the fever cases were diminishing and the tropical ulcers were improving despite the shortage of iodoform and sulphanilamide for treatment.

Rowley rejected the 'spooning out' or scraping out of dead flesh which was the long-established treatment for tropical ulcers among the Dutch and was being adopted by the British. Such drastic treatment damaged and stopped the growth of healthy granulating tissue so that the putrid holes increased in size instead of closing, especially on the shins where there was little flesh, the skin of under-nourished men was tissue-paper thin, and the circulation was poor. His treatment was repeated cleansing of the lesion by bathing, the gentle picking out of dead flesh with forceps and, above all, rest for the affected limb. He also tried the other accepted treatment, by maggots sealed in bandages, but rejected it because the maggots, having eaten the dead flesh, then invaded the granulating tissue.

How to contrive sufficient rest for the sick when the Japanese were putting on the pressure, increasing work hours and demanding more men on work parties, was a problem he had been mulling over for weeks.

He checked Roy Whitecross' ulcer. 'Sure you're keeping up the bathing?'

'Like clockwork, sir.'

'Any other complaints?'

'I've just had a pretty short bout of fever. Felt as if a truck had run over me and my head had a hammer rampaging round inside it. It's over now.'

'All right. See Sergeant Armstrong in the RAP. He'll take out the scrap of dead flesh in that hole. And, Whitecross, wait in the RAP. I'll have a word with you after parade. Next.'

Next was the Dodger.

'What's your trouble?'

'Me balls are red hot,' the Dodger explained graphically.

'Ah! Been trading with the girls in the village too, have you?'

'Me having it off with a boong? No way. I had a bit of a heat rash on the crotch. Put zinc ointment on but the thing's spread. It itches and burns something awful.'

Rowley refrained from asking the Dodger how he had come by zinc ointment. He knew a sharp trader with rupees to spare could buy medicaments from the Jap camp, where it was believed that Red Cross supplies intended for the prisoners of war were held. When the Japs lost this war, the worst charge against them should not be the bashings they dealt out, which were normal procedure in the Nipponese army, but the witholding of medical supplies.

'Let's see,' he told the Dodger.

The entire scrotum was inflamed, an angry red. There was no doubt that the Dodger was suffering from the dermatitis of pellagra and it was severe.

'You'd better start spending your money on fresh fruit instead of cigars and chindegar. You're suffering from a vitamin deficiency disease, pellagra. No Duties. Go and lie down . . . and don't use your contraband zinc ointment. In this climate ointment keeps the skin moist, a breeding ground for fungus and germs.'

The next three cases all presented the same symptoms, pronounced oedema. Not only the ankles but the soft tissues of the face were puffy.

'Reckon it's the beriberi, Doc. We shouldn't be out working on that line. Even the Nips don't expect blokes to work with the beriberi.'

Rowley was thoughtful for a moment. The three were mates and also members of the Dodger's little clique.

'Light Duties, one day, all three of you.'

In their sleeping hut, the three mates converged on the Dodger's area of sleeping space, their spokesman snarling.

'How about that, Dodger? You reckoned we'd get sent back to base 'ospital for a spine bash. What d'we get? One day . . . one day on Light and you're the one that gets to spine bash. Right this minute I could have yer balls.'

'Right this minute you'd be welcome to 'em.' Dodger scratched the parts under discussion gingerly and winced. 'I don't know what gives with that Doc Richards. Sends out sick blokes and lets a lot of bludgers hang around camp.

Doesn't know what he's doing. And plays favourites. See him send Whitecross off to Old Silver's tender loving care just for one measly ulcer? You blokes take care of yourselves. I've got me own worries.'

Dodger fumbled in his kit for the zinc ointment.

Sick parade concluded, Rowley went to have a look at Whitecross, whose leg had been rebandaged. 'You're a good lad with your hands. How would you like to help out in the RAP?'

Roy grinned. 'Suits me, sir. Anything to give a hand.'

'Rowley's a good sport,' he thought as he watched the RMO walk out of the hut.

Rowley, walking across the parade ground, was thinking too. Whitecross is only twenty-two. Give him a chance and he might beat cerebral malaria and pellagra, and he might get to be sixty with two legs to stand on.

The Dodger's mates had gone off grumbling to report to Major John Shaw, in charge of canteen supplies, for Light Duties. The Dodger, moving his legs restlessly, was reading a dictionary, not from any love of learning but because he was training a quiz team on whom he intended to do some heavy betting. He expected to clean up in a few days.

'You are sick. I'm sending you back to base,' Rowley told him.

The Dodger immediately thought of the effect of being away from his carefully established business interests.

'For a rash? I'll be all right after a few days off.'

'Pellagra isn't just a rash. You had a rash and mouth sores at Tavoy. That was the first stage. The second stage progresses from dermatitis to dementia . . . that means, without the aid of the dictionary, that the disease is attacking your brain cells and you go mad. Then you die. My job is to keep you alive if I can. Get your kit together. You're on the next truck that comes through.'

'Well, it'll be a bit of a holiday. Thanks Doc . . . sir.'

'By the way, you've been buying a lot of salt,' Rowley continued casually. 'Are those three mates of yours good customers? Did you sell them salt yesterday?'

'Yeah, I s'pose they are good customers. And yeah, they took all the salt I had yesterday. Why?' Dodger was scowling suspiciously.

Rowley chuckled. 'I bet you told them what else to do.' He strolled out of the sleeping hut as Dodger's expression became a mixture of consternation and amazement.

Don Kerr was attending to the typing of a new nominal roll in triplicate, demanded by Jap headquarters and to be ready as of yesterday.

'I had three cases of fake beriberi this morning,' Rowley told him. 'Self-induced salt oedema. Some of the fellows have dropped to it that you can cause oedema by shovelling down a lot of salt with your food and taking a big drink of water. Bingo, the next day you blow up like a balloon and the Nips, only believing what they can see where sickness is concerned, think there is a severe outbreak of beriberi.'

Don Kerr shook his head wonderingly. 'There is no telling what tricks the lads will think of next.'

'Trouble is, it isn't really funny,' Rowley continued. 'On a minimum protein diet the water stays in the tissues instead of being filtered and passed out of the body. It reaches the lung tissues and with pulmonary oedema they literally drown themselves. We'd better let Sid Krantz know about this and arrange for a strict control of all salt intake.'

Rowley told Colonel Anderson about the salt oedema as they waited for Sid Krantz and Don Kerr to join them for the afternoon bridge game. The Colonel's voice was dry with annoyance when he spoke.

'They are a lot of loafers and whiners. I'll address the troops myself and order a stop to all private purchase of salt.'

'I hope you'll warn them how dangerous it is to induce oedema. They can kill themselves.'

Anderson bent his head over the cards he was shuffling and gave Rowley a quizzical look. 'Surely, Captain Richards, that is exaggerating, perhaps a trifle?'

'Definitely not, sir. The practice is potentially lethal.'

'Many things are potentially lethal in this world, but we survive them. I know these men are good soldiers in battle I've seen them, they fight as well as any and yet . . . now . . . well . . .' Anderson paused. Finally he put his thoughts into words slowly: 'Now they tend to behave like spoiled children. They complain and they shirk. It's immature. Reasoning with the Japanese is difficult at the best of times, but with so many men claiming to be sick the Japanese are becoming very touchy about their work schedules. The most important thing to them is the building of their railway line, whether we like it or not. A little cooperation would make conditions easier for the entire force.'

He smiled. It was a smile rarely given but quite charming, inspiring confidence in the men he commanded.

'While we are on the subject, it should be possible, don't you think, to discourage the large numbers who report sick with trivial complaints.'

'There are large numbers who are sick, sir.'

'Well . . . we shall see what a few pep talks can do to boost morale. And try to get more men working. It's really good for them to have work. We saw that at Tavoy.'

The CO is certainly right about Tavoy, Rowley thought. But Tavoy is not Hlepauk and the aerodrome was nothing like this railway. Here we can't go by the book; it's almost got to be by instinct.

The discussion could not be carried further because Don and Sid arrived ready for some serious bridge. Krantz was an expert contract bridge player and Anderson was very good also. Rowley and Don Kerr were learning. It was an incongruous scene, four men seated at a small table fanning out their cards, considering, making their bids. They could have been in a city club or a Potts Point apartment in Sydney, but their shelter was a wall of living bamboo as confining as barbed wire. Only a kilometre from the camp teams of ragged undernourished men were laying and spiking the iron rails that were reaching out from Thanbyuzayat into the Burmese jungle. But the contract bridge sessions were a necessary therapy, a brief relaxation from tremendous responsibility . . . equivalent to playing tricks on the Korean guards, compiling recipe books when there were no ingredients, or thinking up ways to defeat the Nip work schedule.

The comparative peace of Boxing Day was mildly disturbed during the 1800 hours roll call, or *tenko* (since the Japanese insisted that the prisoners use Japanese terms and numbering). One of the work parties was late and this had made the guards irritable. Until all prisoners were accounted for they could not settle down in their barracks for an evening's relaxation. Rowley was about to start evening sick parade when the guards started routing everyone out for an extra tenko.

'Have you heard what's up?' Rowley asked Don Kerr.

'Tomoto's work party came in one man short. He may have got back to camp on his own.'

The head count was made twice and it was definite that a man was missing, Sergeant Ron O'Donnell. Corporal Tanaka organised search parties of the guards to go into the jungle and the men were ordered to stay in their huts. The officers gathered in the administration office.

'Apparently another attempt at escape,' Anderson said. 'A brave fellow, but foolhardy. One can only hope he may be the one to succeed but he hasn't much chance.'

'O'Donnell isn't a foolhardy type,' said Don Kerr. Like Rowley, the record-keeper Allen Taylor, and Jim Armstrong, Don knew every man in the camp, his history and something of his character. He glanced at Rowley. 'What do you think?'

'O'Donnell has just gone back to the work parties after being on Light Duties with malaria. He's a sensible fellow and would know he isn't fit enough to try an escape.'

'An accident, then,' Anderson hazarded. 'He may be lying hurt somewhere.'

A young soldier tapped on the door lintel and slipped into the administration office.

'What are you doing out of your hut, soldier?' Anderson demanded.

'Sir, the guards aren't around and I had to talk to you or one of the other officers. About O'Donnell.'

'Well then?'

'He was the sergeant in charge of our kumi today. I don't think he'd shoot through. He was always telling us not to try, it couldn't be done. "Peanut" . . . the guard Tomoto . . . took him off into jungle just as we were heading back for camp.'

'You saw them go off?'

'Yes, sir. Tomoto had ordered us back to camp, then he called Ron O'Donnell away. It was something about extra timber for an embankment. I saw Tomoto come in alone after we got back. He went straight to Corporal Tanaka. Some of our blokes have been talking about it in the hut and they reckon they heard shots in the jungle.'

'Think you can show us where you last saw O'Donnell?' Anderson asked.

'Yes, sir.'

'Right. Go back to your hut and wait. I'll be calling for volunteers for a search party.'

The soldier moved away from the door and Anderson snatched up his cap. 'I'm going to seek Tanaka. I'll need you,' he said to Bill Drower. 'Major Krantz, Captain Richards, you may need to be ready for emergency surgery if O'Donnell is brought in wounded.'

Almost an hour passed before Anderson returned to the administration office. His mouth was tight.

'O'Donnell is dead,' he told the waiting officers. 'Shot once in the chest and twice through the head. And there's something damn fishy about the whole thing. Tanaka hardly let me get more than a glimpse of the body for identification before he covered it up again. He's jumpy. He says Tomoto now admits to shooting O'Donnell because he was trying to escape. Supposedly Tomoto was afraid to own up at first because our men have been terrorising him with threats that they were going to accuse all their guards of war crimes after the war.

'But I don't believe the story. Our men who found the body say that he was lying on his back and he had been shot in the chest at close range. He had no gear for an escape bid, only wearing shorts and his hat and carrying his water-bottle. The head wounds were either to finish him off, or insane mutilation of a dead man. It's straight-out murder. An escaping man isn't shot in the chest. Murder . . . but why?'

Major John Shaw, who had said little since joining the group in the administration office, cleared his throat gently. 'Tomoto's inclinations could be the answer, Colonel. Some of the younger boys have noticed that most of the guards have girlfriends in the village but Tomoto seems to have a distinct preference for young boys. He's tried to get friendly with some of our lads too, handing out little delicacies, showing favouritism in the work parties.'

Anderson's eyes had been widening steadily as Shaw was speaking. His jaw muscles bulged and the grating of his teeth was audible.

'Sodomy? A sodomist!' His voice was sharp with outrage. 'You can't mean that Tomoto would assault . . .'

'He wouldn't get as far as assault,' Rowley cut in. 'Not with Ron O'Donnell. I've heard his views on the subject.'

Anderson's face was reddening with anger. 'I'm going all the way with this—to Colonel Nagatomo. No sodomist is going to murder a man of mine and get away with it. Captain Richards, you and I are going right now to demand an examination of O'Donnell's body. I want you as a witness to the nature of the wounds and that they could not have been inflicted on a man who was escaping.'

But Anderson and Rowley were not permitted to re-enter the guard house. Corporal Tanaka, usually reasonable and cooperative, was withdrawn and adamant in his refusal.

'It is not of use for the doctor to be here. He can do nothing. This man is dead; shot attempting escape. Already we have made box. Tomorrow Thanbyuzayat for funeral, Anderson San.'

The guard house door was closed.

Rowley was too angry to sleep. He stared into the darkness, remembering the camp-fire discussions at Tavoy which ranged over a variety of subjects, and one in particular triggered by a joke that had led to a quite in-depth discussion of homosexuality. Maurice Barkley, the serious young intellectual, had reviewed the history of the Platonic Relationship in Ancient Greece . . . the relationship that Plato considered the only real form of love since only men were intellectual equals, women being lesser creatures and suitable only for the breeding of children and care of the home. Oriental races, Maurice had said, had a similar outlook to that of the ancient Greeks. Both Sergeants Lynch and O'Donnell, whose religious faith was an integral part of existence, had taken part in that discussion, refusing to consider the possibility that a third, intermediate sex was part of the natural scheme of creation. Rowley could imagine O'Donnell, blind to all danger, trying to beat the daylights out of Peanut Tomoto before he took that bullet in the chest. What a way to die, and on the day after Christmas! It was up to Colonel Anderson now to see that O'Donnell was avenged.

Colonel Anderson and Major Barraclough accompanied the wooden box to Thanbyuzayat for the burial.

Red Cross prisoner-of-war lettercards were distributed on 27 December by Corporal Tanaka, the first since the prisoners had been allowed to write a letter home from Changi. The cards were printed:

I am interned at the War Prisoners Camp at Moulmein in Burma.
My health is—good—poor—bad.
I have not had any illness.
I—am—have been—in hospital.
I—am—am not—working (pay at . . . per day).
I am with friends . . .

A space of three lines was added for personal message and signature.

Rowley mentioned two Kyogle men, Duncan McNaughton and Hilton Fairburn, who were in the Hlepauk camp. Would his family's anxiety, he wondered, be eased by the bland messages on a printed Japanese card? If only there was some way to give them a more personal assurance that he was really well. He pondered over the card, began writing the address, hesitated and then grinned. The name of their house was 'Cranmer'. Instead Rowley wrote 'Criswell'. One of the family would surely pick that up.

He returned his card to the administration office with the 5¢ charge, hoping fervently that it would get through to his family. Rowley noted that the card gave their location as Moulmein and wondered if the outside Western world knew anything about the plight of white coolies building a jungle railway. They probably believe we're just sitting out the war, he thought. He also noted that the excitement of being given letter cards had damped down the angry talk about Ron O'Donnell's murder. The Japs were cunning bastards indeed.

Anderson and Barraclough returned by truck in the afternoon. Nagatomo had refused to discuss the killing of O'Donnell other than to state that the prisoner had been obviously trying to escape and the guard was to be commended for having done his duty.

'I couldn't press the matter any further, gentlemen. We had no medical evidence as to the nature of the wounds,' Colonel Anderson told his officers. 'We can only warn the men about this particular danger.'

The matter of Sergeant O'Donnell was set aside but not forgotten.

There was news that a party of Dutch prisoners had arrived at Thanbyuzayat; approximately 1,500, of whom 350 had been admitted immediately to hospital, illnesses ranging through severe dysentery, beriberi and various other deficiency and tropical diseases. They had come from Sumatra and Singapore via Rangoon where there had been 173 deaths. News of deaths in large numbers was now becoming commonplace and had little emotional effect. Only news that a mate had died somewhere on the line had meaning. Ten thousand faceless victims couldn't mean as much as a bag of sour rice in the rations or the escape of a miserable ox. A hungry belly knew only its own need.

As a relief from the pressures of the day Rowley began to spend his evenings in making plans for the future, when he would be free and going on with his civilian life as a doctor in private practice. When he thought about himself other than as an RMO doing his job as best he could, he saw a young man of only twenty-six who was literally growing up as a prisoner of war. His adult experience as a civilian was almost nil. His emotional life, as far as women were concerned, was in cold storage for the duration. As a resident at the Mater Hospital and as an officer at Holdsworthy he had played the field, but sidestepped any suggestions of getting serious as he had a strong conviction that a soldier in wartime should not marry, tying some girl into a situation which could become intolerable. He was very glad, now, of that conviction. At least there was no wife back in Australia wondering month after month, for a year and perhaps another year, if she was the widow of Rowland Richards.

There had been dinners at Romano's that were memories of delightful companionship and good fun, light-hearted flirting and of a discreet head waiter who could say, having attended Rowley with a different companion for several nights in a row, 'Captain Richards, it's been quite a long time since we've had the pleasure of seeing you, sir.'

Before he married, Rowley decided, he would need a definite level of income, an established practice and a house. That might not be the way it would be, but it was the way it should be. He also needed a philosophy for living. These were the meditations of a very young man, doggedly determined to maintain his identity and integrity as a human being in a POW camp in disease-ridden jungle. He needed an older, wiser head to share his thoughts and give him guidance, and in Major John Shaw he had found such a friend. Their evening discussions began at Hlepauk on New Year's Eve, 1942, beginning with a discussion of the possible developments around Sydney and the relative merits of country towns and cities as places to settle down and to practise medicine.

John Shaw was forty years of age, a quiet, patient man and a deep thinker. Before the war he had been the Chief Engineer of the Department of Main Roads in New South Wales and during the Malayan Campaign was the Officer Commanding the 2/12 Field Company of Engineers attached to 8 Australian Division. He had received an immediate award of the Distinguished Service Order for his service during action. His knowledge of the cities and country towns of New South Wales was immense and the country choices, he suggested, could be Kyogle or some other small town; Newcastle or Taree as large towns which would reach city status after the war; and the choice of a Sydney suburb could be the seaside suburbs of Seaforth or Cronulla. Rowley decided that the choice would be either Kyogle or Seaforth.

'Well, now you'll need to plan your house,' Shaw suggested.

Rowley was so excited by the thought that he could not sleep and drew up plans for a house which, with suggestions from Shaw, evolved into an L-shape with a sun-porch, a large hallway and a separate lobby and medical wing. The plan was concealed with his prisoner-of-war diary—a four and a half by three inch notebook in the official, Japanese-required, medical records, it was the one place that was nevear searched.

As well as working on his house plans, Rowley read voraciously. He had read most of the books taken from the Tavoy library. These treasures had become a mobile library, carried two or three volumes in the pack of each man.

At first, in Tavoy, the trashy yarns and the sensational—like *Lady Chatterley's Lover*, which was banned in Australia—had been in demand, but their popularity quickly waned. However, books that really occupied the mind were a defence against the dehumanising treatment meted out by the Japanese, even if the need was not consciously recognised. Men who had read comic strips, sporting pages and the sensational muck-raking reporting of the Sunday *Truth* newspaper at home were reading encyclopaedias and competing in quiz competitions.

Rowley read the Bible from Genesis to Revelation and wasn't unique in this feat. He switched as his needs for relaxation and entertainment or for strong mental stimulation required, from the hilarious and satirical prohibition-era comedies of Thorne Smith to histories of philosophy. Although he had always had a good memory for figures and data, but poor recall of the written word, he found he could recall at will long quotations that impressed him.

On paper purloined from the administration office, John Shaw began writing a history of English literature purely from memory, so they battled against Japs and jungle, brutality and decay.

The New Year, 1943, rang in another move for 'Anderson' Force, this time to Tan Yin where 'Williams' Force was already installed. The move, on Sunday morning, 3 January, was a good one. A track wide enough to accommodate trucks had been cut to Tan Yin and the Japanese provided twenty trucks. The first truck arrived only about one and half hours late, all the gear was moved without confusion and the camp site seemed better than Hlepauk. The view of jungle-clad hills, particularly one magnificent bluff that caught the glow of technicolour sunrises and sunsets, was splendid. Instead of the bare ugliness of a hastily cleared camp site, Tan Yin camp was laid out among trees and shrubbery, and the hundred-metre atap and bamboo huts were better built that they had been at Hlepauk. Tan Yin seemed to support the philosophy that any move had to be for the better.

Rowley was hardly off the truck before he was inspecting the camp's hygiene and sanitation arrangements. 'Williams' Force had done a good job of establishing essential services, but the first and most obvious criticism of the Tan Yin site appeared to be a poor water supply. There was no river. Water was drawn from wells by whips—long bamboo poles weighted at one end and balanced on uprights, so that the weight was just sufficient to counterbalance a bamboo pail of water suspended by a rope at the opposite end. The full pails were emptied into a split bamboo trough which carried the water into a sixty gallon wooden vat. It was a mechanism that had been used by peasants in Egypt at the time of the Pharoahs.

There were no facilities for an RAP for 'Anderson' Force, so Rowley improvised in a building near the RHQ merely by opening the kit and setting up the necessary items on the bamboo sleeping deck for the duration of the sick parade, then packing away again. The Japanese commander promised a tented RAP and hospital and three twenty by ten foot tents arrived three days later, which was prompt for the fulfilment of a Japanese promise. They were satisfactory but there were no floorboards and the bare earth was dry and dusty . . . not very hygienic. The floorboards arrived four days after the tents. Not too bad, Rowley thought. Front-line doctors must be carrying on under worse conditions.

Lieutenant-Colonel John Williams, at first meeting, impressed Rowley as a straightforward, uncomplicated and blunt-speaking soldier of considerable courage. He had undergone interrogation in the notorious 'bicycle camp' in Java—with Oriental persuasions such as water torture and thin slivers of bamboo driven under the fingernails—and made no secret of his hatred for all 'slants', particularly Japanese. The medical officer attached to 'Williams' Force was Lieutenant-Colonel Norman Eadie, CO of the 2/2 Casualty Clearing Station, who ran a good hospital and RAP. Tan Yin was the beginning of a close association between 'Anderson' and 'Williams' Forces as they were moved, frequently leapfrogging each other, from construction camp to construction camp along the railway that was to link Moulmein and Bangkok.

In his spare time, Rowley made improvements to his house plans, drew a map of the Seaforth-Manly area with the help of John Shaw, so that he could decide on the best location for his house, and continued their philosophical discussions as well as a course of lectures on the history of architecture prepared by Shaw.

'Everyone should have their own cardinal rules for living,' John Shaw said. 'Maybe I'm an old cynic but mine are: one—expect nobody to act decently; two—expect nothing to be done correctly the first time; three—never trust enthusiasts.'

'I like them,' Rowley agreed, 'so maybe I'm a young cynic.'

He wrote in his small, secret diary: 'My Rules. One—trust nobody, including oneself. Two—believe nothing without confirmation. Three—keep an open mind only when it is impossible to make a decision after a consideration of all available facts.' Then he added: 'Mother's Rule. Trust in God and fear no man.'

By 10 January the numbers of sick were stationary at 200, including about thirty chronic unfits who had returned from the base hospital at Thanbyuzayat. The Dodger was back, classified as fit, his chief complaint being that he had had to leave the scene of many profitable rackets just when he was becoming comfortably established.

The days were glorious, hot and dry, although the nights were becoming very cold. All the clothes a man possessed, plus the issue of a thin Japanese blanket each, were not enough to keep the awareness of bone-aching chill out of sleep.

The sunrises and sunsets and the ever-changing effects of sun and shade on the hills were spectacular. Tan Yin seemed like a movieland jungle paradise. There was no immediate indication that the prisoners had already passed through the gates into Hell and the making of black history; that God was looking the other way and the only person one could trust was oneself.

7

The Expendables

Colonel Anderson's attitude since both the work pressure from the Nips and the numbers of sick increased at Tan[illegible] had Rowley puzzled. He would have liked to have been able to talk to John Shaw about it and ask for some guidance, but discussing the CO with another officer went against the grain.

Charles Anderson was intelligent, tough physically and undoubtedly physically courageous. He was a natural leader of men, able to maintain discipline and inspire efforts almost to the limits of human endurance. In fact, one of his favourite stories about Africa, where he had been a planter and a leader of safaris, concerned a safari on which the party had to trek through hard, hot, waterless country.

'We tramped every mile of the way to the next plantation, over stony ground and through thorn scrub, with our feet rubbed to a raw, red mush and the blood oozing out of our boots. There's no such thing as quitting. As long as there is life in a man he can keep going.' That was Anderson's philosophy. His attitude to malaria was as tough: malaria was a minor ailment and if a planter in Africa took to his bed every time he had a dose of fever, nothing would ever have been achieved.

Boots full of blood. Limits of human endurance. Was that an example of the same heroic spirit that was keeping the British going through the saturation bombing of London? Or was it a failure to recognise the real point at which the human body did reach the limits of endurance and begin to break down? Anderson detested malingerers, as he had said more than once, and since Christmas at Hlepauk he had suggested on several occasions that there were too many shirkers on sick parade. But was he a competent judge?

'No!' Rowley said out loud.

'Sir?'

Jim Armstrong was studying him with concern.

'It's all right, Jim. I'm not jungle happy. I was thinking about the bloody Nip blitzes on sick parade. How are the lads feeling about it?'

'Damn low. Morale in the camp would be at rock bottom now. They're as surly as bears. And they're not just blaming the Nips for sending the sick out to work. They're beginning to say that officers who don't have to shovel and carry on the line have no right to order sick men out. They reckon that

Major Fisher's visit with the Butcher won't do any good, that the doctors from base don't know what it's like out here.'

'Major Fisher knows now and I think we might be able to ease things a bit,' Rowley told him, but privately he admitted to himself that the doctors from base did not know.

The pellagra epidemic had started in mid-January with an increase in sore throats and swollen, shiny tongues. Within a few days there were 317 sick including 60 with URTI and 100 with early pellagra symptoms. By 23 January 1943, there were 399 sick out of a camp strength of about 500, and 250 of them were showing various symptoms of pellagra. Many of the earlier cases were now suffering from diarrhoea and rashes on the shins and calves that were aggravated by the heat of the camp fires about which the men huddled for warmth in the bitterly cold jungle nights.

One of the Korean guards had appeared at morning sick parade, demanding men for the work parties. He strode down the long lines of men, jabbing a finger at everyone not obviously incapacitated by leg ulcers or the oedema of beriberi and ordered 147 to report for work. Rowley began bargaining with him over each man.

'*Takusan byoki, shigoto nai yon oaio.*' (Very sick, no work five days.) Rowley held up his hand with the five fingers spread.

'*Ichi*,' the guard held up one finger.

'*Ni*,' Rowley said emphatically and held up two fingers.

Grudgingly the Nip agreed. It was like haggling in a Singapore bazaar. It was exhausting but a total victory . . . for that day at least. None of the very sick men was forced out to work, but to save face the Nip ordered him to hold sick parades in the afternoons as well as the dawn parade.

Well, that's the end of my contract bridge lessons, Rowley thought in a flash of frivolity that was a blessed relief after the confrontation.

But that had not been the end of it, even for one day.

The CO sent for him early in the afternoon. Anderson was looking grim.

'You kept 401 men in camp today as sick,' he said without preamble. 'I have just come from an interview with Lieutenant Yamada. He is very angry. This camp provided less than one hundred workers today. And our work record has been steadily declining since we arrived here. If this malingering goes on we can't expect to deal reasonably with the Japanese Command and obtain necessary concessions for the really serious cases.'

'None of the men I kept in camp is a malingerer, sir. We had an epidemic of pellagra.'

'Pellagra! Sore throats and a few cold sores on the lips. We saw that during the cold season at Tavoy. They cleared up in a few days.'

'Only an apparent remission. Pellagra is a vitamin deficiency disease that attacks repeatedly if the conditions causing it aren't properly treated.'

'Vitamin deficiency . . . a lot of new fangled medical jargon. You doctors have as many fads and fashions as women with their clothes.'

'Pellagra was recognised as a disease as long ago as the Crimean War, probably even the Napoleonic Wars, although medical science hadn't progressed to the stage where it could give a name to the cause,' Rowley argued. 'Believe me, Colonel, pellagra, untreated, can be a killer. And the last stages aren't pleasant . . . dementia comes before death.'

Anderson looked uncertain. 'Well, we have to clear up this epidemic. Not only this Force but all the prisoners along the line will suffer if we can't seem to be cooperating.'

'Better quality rations, fresh vegetables, an increase in protein by way of eggs and meat, shorter working hours and one full rest day a week instead of one day in ten.'

'Good grief, do you want the moon as well?' Anderson exploded.

'The moon would be quite useless in the treatment of pellagra, Colonel.'

Anderson smiled his charming, reasonable smile.

'Well, at least we keep our sense of humour, Captain Richards. Look here, Rowley, you know I am continually making representations to Nagatomo for an improvement in the rations and I am lodging complaints about the workload. I believe we will succeed in getting better supplies when the Japanese solve the transport problem. Right now they have worn out old trucks breaking down on the jungle tracks. When the line is laid right up to this camp it should be possible to bring in supplies by rail. But first the line has to be laid. So we need workers.

'The other point is that Nagatomo sticks to his original statement that only those who work will be fed and paid. No work, no food. Can you find some way of getting enough men back to work to give me a reasonable starting point to bargain with him?'

'The only immediate treatment would be to provide some form of vitamin B6. Brewer's yeast is one rich source. We don't have that and I doubt the Japs could supply it. But I have picked up a recipe for a type of wild yeast from Lieutenant Boundy of the Engineers. The culture is made from fermenting potatoes and flour and fed with sugar and rice water. If we could get the ingredients as extra to our ration . . .'

'Right, I'll arrange it with Yamada,' Anderson said briskly. 'I know I can count on your cooperation, Rowley.'

The ingredients for the yeast arrived before evening. The potatoes were boiled to a mash which was then diluted with an equal quantity of water. Rice flour and sugar were added and the mush put into bottles and corked with pads of cotton wool so that the gas could escape. When the yeast developed it had to be fed daily with sugar and starch (mashed potatoes). The juice of half a coconut per gallon every two days kept the mixture sweet. After five days it could be decanted for use. Rowley calculated that he would need twenty-five gallons a day for 'Anderson' Force so that every man could receive one to two ounces daily with a double dose for the existing pellagra cases.

Until the yeast was ready, he would have to do his best to protect the sick yet increase the numbers available for work, so that the Nips would not harass the CO. The long sick parades were tiring and he had developed a cold as well as a mild case of diarrhoea. But he thought it was carelessness rather than exhaustion that caused him to slip on some steps and, trying to save himself from a fall, tear the palm of his right hand on a piece of wire. The cut should have been stitched but he decided that he could not spare the suture material so simply taped the edges of the tear together and had Jim Armstrong bandage the hand.

'But surely it needs suturing,' Jim protested.

'No way. We need the sutures for serious accidents.' Rowley grinned at his medical sergeant. 'Truth is, Jim, I couldn't face directing you, in cold blood, how to put the stitches in . . . and without a local.'

The hand throbbed and he felt almost as surly as so many of the men on sick parade looked. There was no time for niceties. This was production-line medicine. He sat at a table with Sergeant Allen Taylor acting as clerk on one side of him, armed with the nominal roll, and Jim Armstrong on the other with a note pad. Taylor hardly needed the roll. He had a phenomenal memory and knew the medical history of the men almost as well as Rowley and Jim. Taylor called the name, the soldier named his current symptom . . . fever, eyesight bad, rashes, pins and needles in feet and legs, dysentery . . . Rowley prescribed the treatment as rapidly as the symptom was stated, and Jim made a note and carried it out.

Rowley continued to insist that sick parade was a parade, and thoroughly disciplined. The men assembled on time, grouped into their kumis and lined up in sequence according to their Jap POW Camp numbers so there was no hunting backwards and forwards through the nominal roll. Each man was dealt with in order, quickly and efficiently. It was one of the small, important examples of continuing organisation that maintained morale, made it possible for Rowley's team to cope with the ever-increasing numbers of the sick, and saved weakened men from the exhausting waits for treatment, with some collapsing as they waited, that Rowley had seen in other line camps.

The men appreciated the discipline and cooperated. Bill Dykes, a little sergeant who had been an SP bookie in civvie life, was a self-appointed 'Sick Parade Orderly Sergeant'. He beat on his dixie with his spoon and rattled off a barker's spiel: 'Line up there, folks, for the sick parade lottery. Special prizes all this week. Three day's Light for diarrhoea.'

The laughter and ribald bantering were good medicine in themselves.

Rowley knew he was sending the mild sick out to slave on the line, but by sending different men out each day he could ensure that all had a little extra rest.

Still, the larger work parties were not large enough to satisfy the Nips. They blitzed the morning and afternoon sick parades and Rowley was terse. Between the cold and his throbbing hand, he did not feel placatory. The Nip guards brought the CO to check the sick parades, and he was diplomatic and reasonable, but Rowley found that Anderson was bargaining to send men out, not to keep them in camp. Unfortunately Anderson did not seem able to comprehend illness. If a man wasn't actually doubled up with cramps or shaking with fever, out he went. A small leg ulcer was not serious.

In several cases Rowley found he had to take a stand against his CO. It was one hell of a situation.

The yeast was ready for use and the result was good. Pellagra cases in the early stages began to show improvement, the sick figures dropping by about ten a day, but the advanced cases, especially those with diarrhoea, were still in a bad way. There were rows with the Nips over every sick parade and Kumi 43 was blitzed. In private, discussions with the CO were tense and almost arguments, but at least Anderson was persuaded to press Lieutenant Yamada for an arrangement that the men would work on alternate days and the sick would do 0.8 metres of earth-moving—half a day's work.

On the day after the arrangement was put into practice, the camp had had visitors from Thanbyuzayat: Colonel Nagatomo, Lieutenant Higuchi, Brigadier Varley and Major Ted Fisher of the CCS who was then in charge of the base hospital. Major Fisher, by some miracle wearing an unpatched, immaculate uniform, made a point of inspecting the pellagra cases, asking Rowley for his diagnosis and opinions. Then the visitors went in to a conference with Anderson, Sid Krantz and John Williams. As Fisher was leaving, he stepped briskly into Rowley's RAP.

'Nagatomo has agreed to meat and fresh vegetables being purchased locally from the Burmese. It should help. I consider you have a problem here.'

The next day, at afternoon sick parade, Don handed Rowley the carbons of a couple of typed sheets.

'It's a report Ted Fisher made for the Brigadier. I thought you'd like a copy.'

> Report on Pellagra at Tan Yin Camp: I have examined samples of patients diagnosed as Pellagra at Tan Yin Camp provided by Lt. Col. Eadie (8) Maj. Krantz (38) and Capt. Richards (50). These men complained of the following signs; cracks at the corners of the mouth, smooth sore tongue, sore lips, skin pigmentations, roughening and loss of weight and complained of the following symptoms: sore throat, salivation, pains in the feet, loss of appetite, diarrhoea, weakness.
>
> I saw no case of recent acute skin rash. The MOs consider that this is because they have learnt to recognise the earlier manifestations such as sore throats, sore tongues and salivation and have put the men off duty before skin conditions developed.
>
> This is probably part of the explanation but in addition it must be remembered that skin changes do not appear in all patients; similarly all the above signs and symptoms were not present in every patient.
>
> I have no doubt that the men I saw were suffering from Pellagra.
>
> Pellagra is a general constitutional disease producing weakness of which the above symptoms and signs are only external evidence. It is due to deficient diet—the deficiencies concerned being protein and vitamin B6 (nicotinic acid). As is commonly seen in peacetime, it is a seasonal condition, getting worse and then better and slowly progressing down hill in successive seasons.
>
> In POW conditions as instanced in Turkish prisoners in the last war,

it is a steadily progressive disorder which can produce a death rate of 16%.

Pellagra can be cured by nicotinic acid which should be provided in tablets of 50 milligrammes to be taken by mouth, for instance, three to six per day, and in ampoules for injections, 100 milligrammes for acute cases. Under present circumstances, this is not available and in any case, such specific cure does not touch the problem of the underlying causal deficiency and therefore will not prevent the occurrence of new cases.

The only other methods of treatment are: (1) the provision of Vitamin B6 in food—eggs, milk, brewer's yeast, and to a lesser extent other foods containing the Vitamin B complex, such as the fish *Kachang Hijau*: and, (2) an increase in proteins.

With regard to B6 containing foods, it must be pointed out that the best dietary supplements we can hope for from available supplies will only provide a bare sufficiency but in time that this should be sufficient to bring about recovery in the milder cases who are in the great majority. I have seen this happen repeatedly at Changi, Tavoy and Thanbyuzayat.

With regard to (2) it is necessary to understand the principle of 'available proteins' established in the 1914–1918 War. That is to say, the total amount of proteins that a man eats is not the amount available to cure his disease.

Two factors affect it: (1) the effect of previous diseases such as dysentery, which impairs absorption; and, (2) the effects of work which use up more proteins. It has been observed that men on the borderline will show no signs of Pellagra at rest but develop them as soon as their exercises are increased.

From the last consideration it will be seen that a certain *number of men should be able to work for half a day who would show signs of disease on a full day's work*, and further that *many men could manage 1.2 metres who could not do 1.6 or 1.9.*

Conversely, *if the remaining men who have not shown signs of Pellagra are given an increased amount of work they will be likely to develop the disease.*

Recommendations: 1. Supplies of Nicotinic Acid. If Dr Higuchi cannot obtain this as he fears, the only source of supply would be a Red Cross ship. 2. Canteen supplies to all camps equivalent to those in Thanbyuzayat Base Camp. If this can be achieved I feel sure that Pellagra as an epidemic disease will disappear and that only a few cases will occur either so mild that they can be dealt with in the camp hospitals within two weeks or so severe that they should be transferred to the base hospital.

(Sgd). W.E. Fisher. Major. 11.2.43.

Rowley laid the report on his table and stared up at the grub-riddled atap of the RAP roof.

'Thank God for that,' he said aloud. 'Ted Fisher is actually supporting us.'

And that was something. Fisher didn't agree with anyone else's opinions out of the goodness of his heart. He liked to arrive at his own conclusions and was a fair bastard to convince.

But Major Fisher's report confirmed Rowley's view of the seriousness of the epidemic. And a copy of the report had gone over to the CO. Maybe it would ease the tension and there would be cooperation. It was hard enough battling the shortage of supplies, the jungle and the Nips without being at cross-purposes with one's commanding officer . . . and trying to conceal from the men that friction existed.

So Rowley's reaction when he read the report was one of deep relief. Fisher agreed with the pellagra diagnosis and the report, considering that nicotinic acid was unobtainable—Dr Higuchi's 'fears' would undoubtedly be well-founded and Red Cross ships were just pipe dreams—boiled down to an opinion that the men must have more protein and the workload must be decreased.

Rowley was summoned to the CO's office and went in optimistic mood. Now that Anderson had a senior medical officer's report on pellagra in the camp, they would be able to cooperate for the good of their Force.

Anderson was weary. As the most senior officer he had the responsibility for the administration of the entire camp at Tan Yin, 'Anderson' and 'Williams' Forces combined, 1,700 men. And the problems of 'Black' Force along the line eventually came to him as requests for assistance or repercussions from the Japanese. He made frequent bone-jolting trips to Thanbyuzayat for conferences with Brigadier Varley who was, it seemed, the only high-ranking officer left in the entire theatre of operations. Word had filtered along the jungle grapevine from Changi that all top-ranking officers had been sent away to Korea. A policy, it seemed, of destroy the body by cutting off the head.

Thanbyuzayat meant conferences with Varley and interviews with Nagatomo, the arrogant overlord of No.3 POW Camp in Burma who aired his intellectual superiority by conducting most conferences in French and could change in an instant from courteous, almost friendly negotiations to rage and a scornful refusal to consider any request. He had to be tempted, cajoled and lured towards every concession.

The burdens of such diplomacy fell on Anderson. Varley was often damagingly blunt, although he had evolved methods of wheeling and dealing with Nagatomo. And the officers on the line seemed to have no conception, Anderson thought, of what he was trying to accomplish in his negotiations with the Japanese.

There had been an occasion a couple of weeks before when Anderson had visited the railway works where kumis from 'Williams' Force were building a bridge over a gully. Pressure was on from the Japanese engineers since the Allied forces had stepped up their bombing. Flying Fortresses had plastered Rangoon and Moulmein. From the secret camp radio, the senior officers knew the sweeping advance of the Japanese down the islands of the Pacific towards Australia had been halted in New Guinea, on Australia's doorstep, where the AIF was embroiled in desperate jungle warfare to keep the foreign invader from reaching their shores. The Japs, beaten back by Aussies, had already suffered their first land defeat at Milne Bay. In such circumstances the only way of getting concessions from Nagatomo and the engineers was to cooperate with them as far as honour permitted to gain their cooperation. On the occasion of the line inspection, Anderson had promised the Japanese a good day's work from all kumis. Just when it seemed that the Japs were placated and in a reasonable mood, Colonel Williams had made a point of getting 'smokos' for the prisoners: ten minutes break each hour with a two-hour meal period. Williams was stubborn and exercised little tact with Japanese.

Corporal Tanaka had lost his temper and shouted that the railway people could work the prisoners twelve hours a day if they liked. According to Bill Drower, one of the Japs from the railway headquarters had said to Tanaka, 'Why should we argue and be polite to them. They are all only dirty white prisoners.'

Yet John Williams could not seem to understand that his obstructionist attitude only bought more trouble than it prevented. As they tramped back along a narrow foot track in the almost complete darkness of the jungle, he had protested against Anderson's promises of a good day's work.

'They don't play fair,' he had argued. 'They set us tasks for the day so the men really pitched in and got the work over by mid-afternoon so that they could have a few hours of relaxation in daylight. What did the Japs do? Next day they increased the work load, reasoning that if the men were able to finish early, then the daily tasks set them had been insufficient.'

Anderson was irritable. 'Then perhaps the men are capable of doing more. Your Australian troops are as fine fighting men as any in the world, but in civilian life, with their unions and all that, Australians have rather a reputation as dodgers of work . . . "Swinging the lead" as you say.'

'Damn it, you've just seen them. Underfed scarecrows pulling on a rope to work a pile driver, manhandling ton beams up to a height of twenty feet. They work like coolies, like bloody slaves on the pyramids. It's primitive—prehistoric!'

'What do you expect? Sophisticated machinery in a jungle? They're lucky to have elephants.'

'And the elephants drop dead of overwork in three months,' Williams had growled.

Unfortunately Williams made no attempt to conceal his attitude from his men and they took their cue from him in dealings with the guards. A man had been severely bashed on the line. Reason? He had refused a guard a drink of water. The water wasn't sufficient for the men as it was, Williams had said, and had insisted on making a written report. Such demonstrations of ill-will in petty matters could do no good at all.

Corporal Tanaka had retaliated by dragging the CO out of bed to stop the men singing in their huts at night. He said they would not be fit for a day's work if they sang half the night.

It was all very difficult.

The young RMO of 'Anderson' Force was another difficult officer, the CO thought. A lad of only twenty-six, he was chock full of theories on hygiene and dietary deficiencies, coddling the men and currently kicking up a hell of a fuss over the complaint that Williams rather graphically called 'foot and mouth disease'. Richards kept insisting that pellagra, which was quite common in armies, was a killer. But there was really nothing to support his assertions. Major Fisher's report certainly confirmed the pellagra diagnosis, but stated that with adequate canteen supplies the epidemic would disappear, mild cases could be dealt with in the camp hospitals and the severe cases should be transferred to base hospital for treatment. It was significant that Major Krantz had submitted thirty-eight cases for inspection and Colonel Eadie had been able to supply only eight from 'Williams' Force, while Richards managed to produce fifty. Of course he was exaggerating.

The interview with Anderson dashed Rowley's hopes of understanding and cooperation. The CO had obviously fixed on the more sanguine comments in Ted Fisher's report, failed to understand the comments on 'available proteins' or the heavily stressed observations that men on part-time work would not show signs of the disease but would break down if worked a full day, and that fit men would develop the disease if their workload was increased.

The argument over the report, seen from diametrically opposite points of view, became heated and both CO and RMO lost their tempers.

'Sickness, nonsense! They're malingering. I've had men on safari in Africa who could go for weeks of sustained effort on jerky and water. They battled through without getting sick.'

'Could you have judged the damage?' Rowley demanded. 'You are not a medical man, sir, and native bearers are conditioned to a completely different diet from our men. In this situation I am the specialist on the spot. If I am to take responsibility for the health of the men then my opinion should be given serious consideration.'

'A specialist's opinion should be queried and not necessarily accepted,' Anderson said dryly. 'You doctors bury your mistakes.'

'As long as I don't have to bury yours,' Rowley flared. 'One of us has to accept responsibility. If you choose to act against my advice, I must apply for a transfer.'

Anderson simmered into patient reasonableness. 'My concern for the welfare of the men is as great as your own, Captain. But we are looking at this thing from different angles and I wish you would try to understand mine. You are very offhand in the way you run those production-line sick parades, which isn't the way to engender the confidence of other officers. You could, perhaps, develop a little more . . . hmmm . . . bedside manner. And to try to put on a little showmanship for the Japanese. They are showmen themselves and appreciate the art in others. We can be tactful, put on a show of cooperating. Given time I can educate them into understanding the problems of white men working in jungle conditions.'

'And given time, sir, I can beat this pellagra epidemic. If we get proper rations and the workload is eased.'

'Believe me, I shall be doing my best,' Anderson said sincerely.

Through February, March and April 1943, as the strength of the Allied counter-offensive increased, the Japanese treatment of the prisoners deteriorated. The railway engineers had only one objective, to complete the Burma-Thailand railway and get their rolling stock moving. The Imperial Japanese Army needed transport for troops and supplies. In the achievement of that object all workers, Burmese coolies and prisoners of war, were expendable. Work—or no pay, no food. Work hours were increased. Reveille was before dawn, sick parades were held in the chill darkness by the light of flares—high-piled camp fires. Work parties were on the line by 0830 hours and did not return before 2100 hours. Water for washing was in short supply and exhausted men stretched out on their narrow allotments of the bamboo sleeping platforms with the dirt of the embankment building work still coating their bodies. It was the first time they had failed to wash and shave. They no longer put on extra clothes to go to bed. They no longer had spare clothing. Thin blankets were wrapped around almost bare bodies and in the darkness men coughed all night.

The Japanese economics officer from Thanbyuzayat visited the camps on a clothing issue inspection. The men were warned that if a man obtained an issue of clothing while still having clothing in his possession he would be punished. A pullover or a service jacket counted as a shirt. Men with any sort of clothing at all were denied an issue and told to mend what they had. But patching material, cotton and needles were like gold and the fabric of shorts and shirts so rotten that stitches would not hold. One man's shorts, John Williams observed, were so ragged that they looked like a grass skirt, but he was refused a replacement. Japanese rubber shoes drew the feet unprotected by socks, were perished, and wore out in a week. So the men worked barefooted, their feet becoming hardened and calloused while the Dry lasted.

The work fever of the engineers became a frenzy by the time it passed—with screamed demands, reprimands and slaps—down to the Korean guards. Beating became more frequent and more brutal. Lieutenant Yamada was shuttling between Tan Yin and Thanbyuzayat, trying to obtain canteen supplies and during his absences the guards really turned on the bashings. One man was bashed by two guards in camp after his work party had returned. When Williams tried to intervene he was told that the man had been 'very bad'; when struck by a guard during the day he had adopted a 'threatening attitude'. So the two guards took turns in slapping him until he learned to keep his hands still at his sides when slapped.

Equipment was almost sacred to the Japanese and when a pick was found discarded at one work site the ganger systematically beat twenty-seven men of the offending kumi so that he would be sure of chastising the culprit. He belted them about the head, neck and shoulders with a stick and one man was knocked unconscious.

Lieutenant Yamada returned from Thanbyuzayat looking very old. He had been relieved of his duties on the line. The prisoners did not know why but suspected he had been put on the mat for not getting enough work out of them. The bombing of Rangoon by the Allies, Yamada told Williams, had been severe. John Williams was sorry to see him go as he had been quite decent. It was to be hoped that the new man would be helpful.

In fact Lieutenant Hasoda attempted a friendly start as camp commander. He put on a party, commandeering the camp orchestra—some remnants of the 22 Brigade Band—and inviting Anderson, Williams and Bill Drower to drink brandy with him. John Williams did not touch his brandy. He had strong views on any form of fraternisation with the 'slants' as his memories of interrogation camp tortures were undimmed. Anderson sipped his and attempted to discuss the camp problems, but was told they would be dealt with 'tomorrow'. The orchestra played and the men sang 'They'll Be Dropping Thousand Pounders When They Come' and instead of the 'Ai, Ai, Yippee Yippee Ai' chorus, substituted 'Ni, Ni, Nippee Nippee Ni'.

Rowley hoped that Hasoda would have had too much brandy to understand the substitution. The new commander was not actively aggressive and it would be better to keep him that way. Rowley was working on 'showmanship' and doing his best not to tweak the tail of the Oriental devil.

Having to deal with Corporal Yamamoto's tail, however, gave Rowley a chance to play the devil a little himself, without being detected. Not having access to the 'hors de combat' provided for the Japanese engineers, Yamamoto had patronised ladies in the Burmese villages and developed a syphilitic ulcer on the leg. He obtained a dose of NAB (the accepted treatment for syphilis) and brought it to Rowley to inject, since he would be punished for having contracted syphilis. Somehow some of the NAB escaped from the vein when the injection was being given, causing considerable pain. Yamamoto obtained morphia, which the Japanese could find for themselves although there was none available for the POW medical corps, and Rowley prescribed hot foments. The hot foments, which had to be continued for several days, were slapped on by Jim Armstrong, very hot.

The line snaked through the jungle towards the Burma-Thailand border. Construction camps were moved into the jungle ahead of the existing construction to dig cuttings, build bridges and make embankments, working back to the last laid line. Then the line layers moved forward. Their camps were ahead of the line, so that supplies still had to be brought to them by ox cart or, less frequently, by the superannuated trucks that were kept running by the ingenuity of Australian mechanics and drivers. Daily the line layers walked back from their camp to meet the flat-top trucks pushed by little locomotives. They unloaded, then laid and spiked the sleepers and rails. The most forward camps were a mixture of British, Dutch and Australian. 'Williams' and 'Anderson' Forces seesawed between construction and line-laying but 'Anderson' Force were the main line layers.

Line-laying work was much more demanding and exhausting than earth moving.

Earthworks could be regulated, tasks could be reduced by subterfuge such as moving pegs, and loads could be reduced and regulated according to the strength of the carriers. But with line laying sleepers and lines were of fixed size and weight which could not be reduced to adjust to the capacity of the workers.

Existence became a confused nightmare of being shunted from camp to camp, sometimes forward, sometimes back towards Thanbyuzayat. Rowley found himself holding sick parades at Tan Yin, the 35-kilo camp, Kunknitkway, the 26-kilo camp, and Anarkwan, the 45-kilo camp. Kunknitkway was a terrible camp. When 'Williams' Force arrived it was still occupied by POWs of the Netherlands East Indies Army who were due to move out to a forward construction camp.

Australian and Dutch were packed into accommodation which was insufficient even for the Dutch. The huts were in bad repair with sections of worm-riddled atap blown from the rafters, bug-ridden and generally dirty. Rats and fleas vied with the bugs to torment the flesh. Cattleyards were next to sleeping accommodation, there were no latrines, but there were swarms of flies. John Williams sent one of his best pilferers to acquire a pick and shovel, which were successfully pinched from the railway equipment store, and set the men who had not been taken along the line to digging latrine holes and refuse pits. There was no wood to make covers and no disinfectant.

Within four days, one hundred of 'Williams' Force were sick with amoebic dysentery.

Sid Krantz and Lieutenant-Colonel Eadie, whose health was failing under jungle camp conditions and the continual nervous stain of dealing with the Korean guards, were detached from 'Anderson' and 'Williams' Forces and sent to the 30-kilo camp, Retpu where a permanent hospital was to be established. They were to take all the sick from 'Anderson' Force with them so that only fit men would move to join 'Williams' Force at Kunknitkway.

There were so many sick, everyone in the camp had some complaint—URTI, tropical ulcers, dengue fever, malaria, pellagra of varying severity, diarrhoea, amoebic dysentery—that the decision on who should go to the hospital camp was very simply made. All men off duty on moving day were evacuated as hospital cases. Some were bed-ridden, some were on No Duties, some with minor ailments were on Light Duties and some were just having their unofficial day off.

The day off had been devised by Rowley and Don Kerr to rest the sick in rotation, while satisfying the CO that they were providing sufficient men for the work parties to keep the Nips quiet, if not happy. Each night Don and Rowley went through the work sheets checking which men had been working consistently and had not had a break for several days. They were marked sick for one day and a corresponding number of convalescents sent out to replace them. The replacements were then reviewed on the next day's sick parade and put back on Light Duties. It was the only way they could conserve the energies of the Force.

The heat and Wet season were with them again. Dust had turned to mud and the fetid odours of the jungle were almost tangible. In pre-dawn darkness and driving rain Sid Krantz, Eadie and the Padre, Pessimistic Pete, set out with 130 sick to tramp the five kilometres back to Retpu. They had no means of lighting the darkness of the track and stumbled and slid through the mud, light sick carrying the stretcher cases or supporting mates with tropical ulcers who were able to limp along.

Rowley prepared another twenty-five to be evacuated the following day and sent them off, escorted by Pinky Rhodes, Arthur Baker and Don Booth. Jim Armstrong remained to help pack up the RAP.

'Now that we have sent the sick to hospital, the Nip attitude will be that all the remaining men are fit and will be fit for all time,' Rowley observed to Old Silver. He grinned but without mirth.

The remainder of 'Anderson' Force, carrying their gear, set out on foot in the wake of the last twenty-five evacuees. Rowley and Jim Armstrong carried the heavy medical packs on their backs. Both were suffering from diarrhoea, but as the gripes caught them only five or six times a day, Rowley classed it as a mild inconvenience.

They passed through Retpu on the ten-kilometre tramp to Kunknitkway. That was to be the forward hospital for the severe sick since it was no longer practicable, according to the Nips, to transport them all the way back to Thanbyuzayat. Hospital! A collection of bamboo and atap huts where Sid Krantz would have to struggle with overcrowding and shortage of medical supplies and equipment. For all that Sid can do it'll be a hell hospital, Rowley thought. Any men I sent to Retpu might as well take their death certificates with them.

The Dutch had just moved out when 'Anderson' Force arrived at Kunknitkway. John Williams and his sanitation construction team laboured from daylight until dark yet the filth and the flies were still appalling. Rowley's hygiene-conscious mind was in revolt. He saw the CO. He supported Williams' requests for timber, flyproofing and disinfectants. Once his orderlies arrived from Retpu he set them to cleaning out a section of a hut for the RAP and carrying on the daily skin inspections that had never been omitted since his regiment had assembled at Holdsworthy. He personally checked the sterilisation of eating utensils, although there was little to eat; the rations for the Australians had been delivered to the camp before Williams arrived and had been eaten by the previous residents. But for their canteen supplies, the combined 'Anderson-Williams' Force would have been foodless for three days before the next rations arrived.

The fit men were not fit for long. Rowley held three sick parades a day. He was now the only medical officer with the Force and because of the diarrhoea epidemic, worked from dawn until late into the night. The total sick rose to 51 per cent. Skin complaints—ulcers and the tormenting pellagra rash—rose from nil to 36 per cent.

John Williams began damming a small river to provide a bathing facility. Railway work was line-laying and ballast-laying. There were no rest days, no concerts.

Rowley and John Shaw continued their mental stimulation sessions at night by their small cooking fire. They had formed a survival team of two, pooling their meagre rations. There was no meat for several days and the 800 grams of rice per man was supplemented by dried beans. A few eggs could be bought from the Burmese traders, using canteen funds, but the price was a rupee for eight eggs, so two cost a man's pay for one day's work. Some enterprising pilferer managed to steal oatmeal from the Japanese supply store and was selling it in the camp. Rowley and John used their money to buy a supply and luxuriated in real oatmeal porridge, sprinkled with chindegar, for breakfast. While Rowley held the first sick parade of each day, Shaw acted as cook stirring their tin of porridge over a little fire.

'I'm having to revise my theories on survival,' Rowley told John as they were eating their evening meal of rice and watery vegetable stew, meatless because the promised supply of pathetic little Burmese cattle had not arrived. 'I used to see it in terms of the regiment as a fighting force—discipline, camp hygiene, gun and safety drill, like wearing tin hats and carrying gas masks. Rations had to be adequate and nourishing, prepared with strict attention to cleanliness. Physical fitness was a matter of a proper balance between work and recreation . . . sport, concerts, mental exercise . . . remember Changi University? And of course, daily attention to personal hygiene. Now that's all washed up . . . or most of it. We're down to basic survival. We're fighting the jungle . . . heat, cold, rain, filth, disease, despair. About all I have left in the medical armoury is hygiene and rest . . . and trying to keep up the hopes of the men.'

'You can see it then. Survival of the group depends on the will of the individual to survive,' John said. 'The Force has broken down into little groups of battlers, three or four together or just a couple of mates looking after each other. And no one has the right, here, to censure the methods of survival. Conventional morality doesn't belong. We have scroungers, bludgers, parasites, wheeler-dealers and thieves. You and I buy stolen oatmeal and don't ask about the origins of black-market eggs.'

'Although we know damned well the Dodger and his mates get the goods they sell by toadying to the guards.'

John nodded. 'That sort survive. It's the basic animal cunning of the Id. But there are other sorts of survivors. Men who have conditioned themselves all their lives to meeting challenges, developing physically and mentally so they know their capacity and can always tackle just a little bit more. That's self-discipline. I used to wonder why a chap in the Department of Main Roads climbed mountains. He said that, hanging on a cliff face, he learned to know himself.'

'It's some cliff face we're all on now,' Rowley said. 'But I think you've made an interesting point about stress. Medically, it must be beneficial, taken in small, tolerable amounts. The breakdown occurs when it becomes intolerable strain. When we get out of this war it's a thought I'd like to follow up.'

'Why not start now? We've finished your house plans and exhausted my small knowledge of the history of architecture. You have enough material here to work on theories of stress that could be important to military thinking in the future.'

Rowley felt excited. 'Or applied to business and industry. A whole national health scheme!'

But at Kunknitkway he had no time to work on the new idea. The numbers of sick increased until he was seeing 350 a day and running four sick parades: one before breakfast for new sick, one in the morning for 'Anderson' Force old sick, one in the afternoon for 'Williams' Force old sick, and then one in the evening for the men returned from work.

An Anglo-American Force was moved into the already overcrowded camp. They were commanded by Captain Fitzsimmons. Fortunately they had their own doctor, a Dutchman, Captain Hekking.

Jim Lynch was back on line work, his health, he felt cheerfully, definitely on the upgrade although he weighed only ten stone seven pounds, which was light for a man who stood sixty-nine inches on his bare, calloused feet and was just thirty years old. With lads of twenty-one and twenty-two in the Force, that made him feel rather middle-aged. Although poor Chauncey, always sick but battling on, was forty-eight . . . an old man. Scippy Maher weighed only nine stone ten pounds and Rusty Irwin, plagued by pellagra rash and recurrent bouts of dysentery, was visibly shrinking. Without beriberi oedema their weights would have been even less.

Blue Lennox had been sent back to Thanbyuzayat, chronically unfit with tropical ulcers and pellagra, and Jim, Rusty and Scippy were the sole remnants of the 'old firm' that had stuck together since training days. The 'firm' would have been instanced by Shaw as an ideal survival group. They pooled all resources and talents: when pay ran out one or the other contributed some small treasure from the little store secreted in his kit; Rusty took the chance on trading with the Burmese for eggs, beans, chindegar and cigars; Scippy, the schoolmaster, was an expert pilferer and scrounger as well as an imaginative cook; and Jim, the lawyer, had discovered in himself unexpected talents as a cook and an improviser of utensils from the scraps of metal that Scippy liberated from Jap trucks—he could make anything from an omelette pan to a blade razor. Jim was also the cheerful optimist who insisted on carrying on the good old standards no matter what the conditions; a refreshing tong in the dammed river, a shave and change into clean rags before dinner, and a cigar with burnt rice coffee after dinner. Dinner was baked rice cake or a paper-thin omelette stuffed with little purple beans and, when they could get it, a few slices of onion.

Jim Lynch's health had been up and down. At Tan Yin he had had several bouts of malaria. Rowley had put him on No Duties and dosed him with the precious quinine. Then he had gone back to part day work. Despite his protests, Rowley had invalided him back to Thanbyuzayat where he had a happy reunion with Blue Lennox, drank real coffee at the canteen, ate stew with lots of meat and vegetables, and made a very observant tour of the 'City of Rackets'. For five to ten cents anything from rissoles, pies, patties of rice, tomatoes and garlic, sweet rice-flour cakes, shaves and haircuts, and occasionally, liquor, could be bought. Among the manufacturer-vendors who carted their wares through the huts were a member of the Sydney Stock Exchange and a wealthy grazier, whose relaxation was compiling his recipe book. A Sydney solicitor vended his own brew of ginger beer.

His sick leave over and carrying goodies for 'the firm', Jim had hitched a lift on a truck, but it had dropped him off a couple of kilometres from Tan Yin, and it was the hardest hike he had ever made, carrying a butter box full of tomatoes, a large haversack of onions, a kitbag of gear, cigars and chindegar. 'The firm' had some tasty suppers and some of the provisions lasted through the first lean days at Kunknitkway.

Jim was stricken with diarrhoea in that fly-infested camp and his precarious health took a downturn. Rowley put him back on Light Duties which turned out to be cattle-herding. He could sit under a tree and watch his lean charges graze, feeling sorry for them because their ribs stuck out worse than his did.

As he tended his herd, he saw a wonderful sight—a group of cowboys leading their cattle to the 45-kilo camp. They were barefooted and dressed in the latest Burma Railway fashion, hats and G-strings. They were Frazer Caddy, BA; Jack Hellmrich, university and interstate cricketer; George Clark, Lecturer in Agriculture at Armidale University; and Ron Gaul. Frazer Caddy, wearing his G-string with natural elegance, stopped to greet Jim and pass a few remarks on the weather and the severity of the jungle fires around the camp. He bowed slightly as the cattle details moved on.

He could have been stopping for a few words as he passed my table at Romano's, Jim marvelled.

Roy Whitecross was back on full duties. The poor vision, muscular weakness and pains in the feet that had plagued him seemed to have cleared up, but the small tropical ulcer on his ankle had not yet healed and he bathed it daily. He pooled resources with his mates Mac, Jim and Maurice. Maurice took care of him like a brother. Mac taught him some French and he taught Mac shorthand. He was also teaching shorthand to Major Shaw who insisted on paying him, sometimes a couple of eggs, sometimes a few rupees. Major Shaw also arranged for Sergeant Reg Wilstencroft to mend his dental plate which had finally broken into two pieces. Wilstencroft stapled it together, manufacturing the staples from a ring. Roy treated the repaired plate carefully, taking it out when he ate but wearing it at other times as a cosmetic so he would not look like the original village idiot. There was no knowing how long it would have to last, but Roy believed they would be home for Christmas. Bombs had been dropped at Thanbyuzayat and Moulmein had been pasted, streams of Burmese refugees taking to the jungle. So 'Archie' would soon be coming to the long-awaited rescue—but there was no guarantee that the denture staples would hold until Christmas.

Lieutenant Naito, whose fame had preceded him from Thanbyuzayat, took over from Hasoda. His arrival was spectacular. Rather battered and rather drunk, he came reclining on a wire mattress on the back of a truck, and was carried to his billet, grunting with pain. He was no sooner settled in than he sent his batman for the doctor.

Rowley's first impression was of a drunken mass of inhumanity, writhing and whining with pain. Naito had been stripped off so that the doctor could view the numerous purple and yellow bruises on his afflicted body, particularly around the kidneys and buttocks. Rowley gathered that the rumoured ill-feeling between Naito and Nagatomo—so open that stories of their conflicts had reached the jungle camps—had finally reached a climax. Nagatomo had exiled his difficult Lieutenant to Kunknitkway in disgrace and in a final burst of fury had knocked him down and planted a few hefty kicks where they would hurt most. Rowley prescribed suitably painful treatment but Naito actually seemed grateful. He and his batman believed that vigorous treatment, massage and very hot foments, being painful, had to be efficacious. Therefore Rowley was 'very clever doctor'. Rowley wasn't flattered. He had no wish to be the family doctor to the Jap Commander.

The whole camp wondered what their lot would be, subject to the whims of the capricious, hard-drinking Lieutenant Naito.

The first whim was that all available men should work on the line, with threats that all the sick would be sent to work. The RAP staff would be reduced to Jim Armstrong and Miles of 'Williams' Force, all other orderlies being assigned to line work. Rowley succeeded in having the worst sick evacuated to Retpu, out of Naito's reach.

On 17 April Captain Hekking reported a case of smallpox and further suspects among the British troops. Within a few days there were some suspects among 'Anderson' and 'Williams' Forces as well and Rowley established a makeshift isolation hospital in a tent across a clearing from the huts. The Nips were panicky. Smallpox, the filthy, disfiguring disease, had ravaged the tiny, overpopulated islands of their homeland for centuries and they feared it as a killer next to cholera.

Kunknitkway was the smallpox epidemic camp.

8

Risus sardonicus

For three memorable weeks Lieutenant Naito was clown prince of the construction camp. His heavy drinking was unusual for a Japanese officer, since alcoholism was not one of their vices. For the visit of the great General Sasa, Commander of all Thai POW camps, Naito managed to be sober and made detailed plans for the General's welcome. There was to be a guard of honour of Japanese, and another of POW officers.

Colonel Anderson was convinced the visit would afford him his long-awaited opportunity to discuss the plight of his Force with an intelligent high-ranking officer of the Imperial Japanese Army. He worked hard on a formal list of complaints and requests to be presented to General Sasa after inspection of the POW honour guard, during a scheduled inspection of the camp accommodation and canteen.

General Sasa descended from his vehicle and was greeted by Naito. The Japanese officers formed a phalanx about him. He returned the salute of the Nip guard, was rushed along the rank of POW officers, then whisked off to visit preselected clean points. The canteen had been stocked with supplies from the Japanese camp for the occasion, which were promptly removed later, and the RAP was stocked with medical supplies just as briefly. The dam across the river was a showpiece, as was the soccer field which had been built for the recreation of the prisoners but was used solely by the Nip guards as the prisoners had neither daylight nor energy for the frivolity of soccer.

The whole tour of inspection lasted no more than twenty minutes. The local Japanese staff made sure that Colonel Anderson had no opportunity to give his formal list of complaints to General Sasa or to have any contact with him whatever.

For the rest of his tenure Naito was 'on it'. He harassed his guards so much that all of them frequently took to the bush, whereupon Naito would constitute himself into a search party to look for them. Having rounded them up he would give them rifle drill and training manoeuvres. The highlight was the leading of an attack on a hill adjacent to the camp, on top of which the 'Anderson' Force latrines were situated. Having captured the benjos Naito screamed out three banzais for his Emperor.

'And that military achievement ought to get him the Fifteenth Order of the Flying Kite,' Rowley observed to Jim Armstrong as they headed for the 'isolation hospital' to check the smallpox cases.

'With a recommendation to hari kari in addition I hope,' Jim said. Jim, the tolerant and patient, was severely overworked since he had been deprived of all orderlies except Miles, and had diarrhoea as well.

Naito was not, it seemed, satisfied with having captured the latrines. At three o'clock a summary order arrived at the administration office demanding the presence of Colonels Anderson and Williams and Captains Fitzsimmons, Hekking, Drower and Richards at his office.

'Oh Lord, here comes trouble,' Bill Drower sighed. 'Probably means to behead the lot of us and stick our noodles on posts along the line.'

If beheading was the object, Naito had decorated his office colourfully for the passing of the death sentence. All the tables had been cleared of office impedimenta and were covered with brightly patterned sarongs. His Highness was seated at the head of the longest table. As the POW officers entered he rose unsteadily, gripping the edge of the table. If almond eyes could be said to have an owlish look, Naito's had.

'This is a party and at a party Japanese never discuss business,' he enunciated slowly but far from clearly. 'It is a party and you must enjoy yourselves.' He paused, then added with great solemnity, 'I order you.'

Naito's other guest was Sergeant Shimojo, a recent arrival who seemed to be a fairly quiet type. The prisoners had not yet been able to gauge whether or not he was descended from a long line of bachelors.

Naito clapped his hands in the Oriental manner and minions came running, led by his batman. They set a place for each guest with a glass, plate and spoon. The first course was green peas—tinned variety—which the prisoners had not seen for months. These were followed by pears, peanut toffee, sliced cucumbers, onions short-cooked in oil, then more peas and coffee. The rice bowl was in the centre of the table. Each course was served as a separate delicacy, with breaks for coffee and brandy between each. The batman brought two bottles of brandy which Naito opened with a flourish.

Colonel Anderson drank abstemiously and John Williams, true to his principles, did not touch the contents of his glass. When Naito chided him he managed very neatly to shoot the brandy under his chair.

'Naito's had too much already. If he and Shimojo polish off a bottle each the samurai blade is likely to go snicker snack after all,' Rowley muttered to Drower and Fitzsimmons.

'Then we'd better polish it off for him, chaps,' Drower advised. 'Two bottles will be the ration for this party, I'd say. The stuff's precious.'

The contents of the two bottles were finished rapidly while Naito held forth on the victories of the glorious armies of Nippon, the advantages of the geisha system and its introduction into Australia when that barbarous continent had been captured. He was commencing a lecture on the magnificent engineering feat of the Japanese in building the Burma railway when he discovered the bottles were empty.

Naito clapped his hands and the batman came running with two more bottles of brandy.

Gunner Richards, 1 Artillery Survey Company, 1935.

Lieutenant-Colonel J.C. O'Neill, CO, 2/15 Field Regiment, 8 Division. Courtesy *Gunners in the Jungle.*

The Grand Priory in the British Realm of the Venerable Order of the Hospital of St. John of Jerusalem
AMBULANCE DEPARTMENT

St. John Ambulance Association

Patron: HIS MAJESTY THE KING (Sovereign Head and Patron of the Order)
President: H.R.H. THE DUKE OF GLOUCESTER, K.G., etc. (Grand Prior of the Order)
Director of the Ambulance Department: Major-General SIR PERCIVAL WILKINSON, K.C.M.G., C.B. (Knight of Grace of the Order)

FIRST AID

NEW SOUTH WALES CENTRE

President: SIR KELSO KING, Kt. (Commander of the Order)
General Secretary: Commander R. GRIFFITHS BOWEN (R.A.N. (Rtd.) Commander of the Order)

This is to Certify that JOHN C. CROSS has attended a Course of Instruction at the New South Wales Centre of the St. John Ambulance Association and was Examined by Dr Cpt. M. Truscott (M.B. B.S.) and is Qualified to render "First Aid to the Injured."

NEW SOUTH WALES CENTRE
SYDNEY, 19th March, 19 41.
Record Book Reference 26.

Lecturer
Examiner
General Secretary

THIS CERTIFICATE DOES NOT ENTITLE THE HOLDER TO INSTRUCT.

The St John Ambulance Association First Aid Certificate awarded to John Cross, one of the sixty-seven 2/15 Field Regiment members who attended the First Aid Course in Holdsworthy. Courtesy *Gunners in the Jungle.*

Lieutenant-Colonel J.W. Wright, DFC, ED, with Rowley on leave from Tampin. Courtesy *Gunners in the Jungle.*

A.I.F. BUILDS OWN HOM

CAMP IN BUSH IS MODEL

In little more than a week, young artillerymen have converted a wilderness of scrub on the outskirts of Holdsworthy into a model camp which has amazed experienced Army officers.

So far as is known, it is the first time that an A.I.F. unit in training in New South Wales, starting from "scratch," has built its own home.

So well have the men carried out their task that officers say they would "break their hearts" if they had to move from the camp, which originally had been intended only as temporary quarters.

Working in the early stages in heavy rain, the unit provided all the tradesmen for building the camp, from the preliminary survey to the last piece of plumbing.

It has provided a remarkable example of the adaptability of country youths, of whom the regiment is largely composed.

Usually, a unit transferred to a new camp has quarters ready to occupy on arrival, or, in any case, has some permanent buildings as a nucleus around which to build. This regim had nothing—just an expanse rough, scrub-covered country.

On December 30 the C.O. (Lieu Col. J. C. O'Neill) received tel phone orders to move from Ingl burn to a site on the outskirts Holdsworthy. An advance part moved in on the new site on Janu ary 2 and on January 4 the mai body arrived.

The troops had to rely entirel n their own resources to provide uarters, and the C.O. was amazed the extraordinary wealth of lent that was unearthed at short otice.

His unit produced surveyors to out the camp area—not a peg s driven until every detail had en plotted on plans—carpenters, mbers, drain - makers, road- lders, and bridge-builders.

ld German Earthenware

nce January 4, without interfer- with their training, the men have ted tents to accommodate 30 officers and

Cutting from Sydney Morning Herald, January 1941.

Tent Town – Holdsworthy, 1941.

Improvised laundry – Tent Town, Holdsworthy, 1941. (L-R) Jack Wilson, Peter Hindley, Jim Moir, Wal Pamplin, Cec Beverley, Doug Stanborough, Jim Cross, John Cross, Brian West.

'Tent Town' at Holdsworthy, 1941. Cutting from *Sydney Morning Herald*, January 1941. Courtesy *Gunners in the Jungle*.

2/15 Field Regiment AIF Officers 26 July 1941.
Back Row: Lt R.E. Steele, Lt J.A. Godden, Lt K.R. Thompson, Lt H. Hingst, Lt J.F. Ross, Lt P.H. Playfair, Lt R. Robson, Lt R. Martin, Lt J.D. Bull, Lt R.G. Gibson, Lt A.C.H. Edwards, Lt J.N. Shearer, Lt E.V. Hiley, Lt G.R.B. Knox.
Centre Row: Lt G.R. Watson, Lt J. Flowers, Capt C.G. Filmer, Capt F.R. Anderson, Capt L. Lawson, Lt P.S. Withycombe, Lt R.W. McIver, Lt R.H. Fleming, Capt L. Hellyer, Lt T.C.leM. Walker, Capt F.C. Walker, Capt M.A. Lindgren, Lt R.P.A. Rabett.
Front Row: Lt D.J. Makepeace, Capt W.W. Julius, Capt K.C. Moulton, Maj A.F. Ball, Maj J.W. Wright DFC, LtCol J.C. O'Neill ED, Maj J.S. Workman, Lt S.H. Ward, Capt C.R.B. Richards (AAMC), Lt R.G. Wells (8 Div Sigs).
Inset: Lt R. Scandrett.
Absent: Maj E.A. Meyers, Lt D.M. Carment, Lt R.J. Keating, Capt E.M. Dahl (RAAOC). Courtesy *Gunners in the Jungle*.

The remains of the Gemencheh Bridge, near Tampin, the site of the ambush when the Australians first made contact with the Japanese in Malaya. Post-war photograph.

Major-General C.A. Callaghan reviewing parade of 8 Division in Changi on Anzac Day, 1942. Australian War Memorial # 43928.

Jim Armstrong.

'Pinky' Rhodes on pre-embarkation leave. Courtesy F.J. Rhodes.

Ex-POW Reunion, Wagga Wagga, 1970. Members of 2/15 Field Regiment. *Back Row (l to r)*: Bob Gaydon, Peter Hindley, Harry Holgate, Stan Cregan, Bruce Schofield, Dave Dunlevey, Jack Eaton. *Front Row (l to r)*: Hal Richardson, Don Booth, Don Bertram, 'Scippy' Maher, Arthur Baker, Wal Sheaffe. *Daily Advertiser*, Wagga Wagga.

Don Booth, Medical Orderly, 2/15 Field Regiment, December 1945. Courtesy Mrs W. Booth.

Des Makepeace. Tampin, Malaya, 1941. Courtesy Mrs B. Makepeace.

Roy Whitecross, 8 Division HQ, in Johore Bahru, October 1941, age twenty-one. Courtesy R.H. Whitecross.

Jim Lynch and 'Scippy' Maher on pre-embarkation leave, 1941. Courtesy Mrs J. Maher.

John Shaw, pre-embarkation, 1941. OC, 2/12 Field Company, RAE. Courtesy David Shaw.

Lieutenant-Colonel Charles Anderson, VC, MC. Commander of 'Anderson' Force in Burma and Thailand, 1942–43. Australian War Memorial # 38657.

Brigadier Arthur Varley, MC, Commander of 'A' Force in Burma and Thailand, 1942–43. Australian War Memorial # 38642.

Reunion of 2/10 Field Regiment members with Captain Bill Drower, Brisbane 1947. *From l to r*: Clem England, Don Kerr, Vince Broe, Harry Dickenson, Reg Harris, Bill Drower, Ron Miscamble, Kev Conran, Charlie McKechnie, Keith Gardiner, Wally Lergessner, Tibby Colton, Jack Buckley. *Brisbane Mail.*

Improvised kitchen at Tavoy POW Camp, 1942. In the background is the brick oven for baking bread and in the foreground are three 44-gallon drums and two 'tin hat' quarries for cooking rice and stews. Australian War Memorial # 3443.

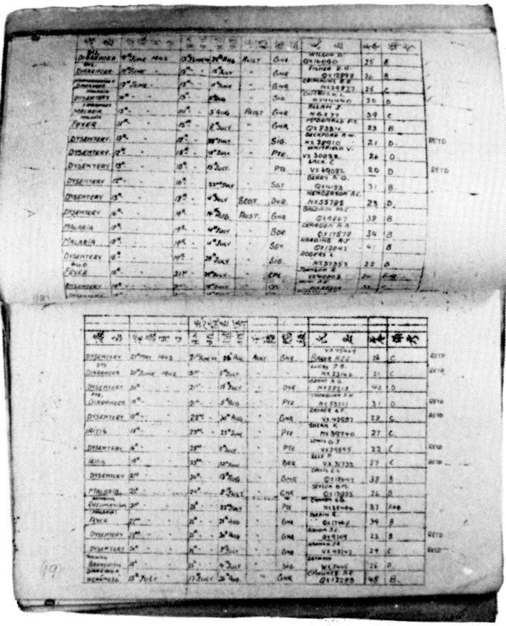

Extracts from the roll of entries into the 'hospital' at Tavoy in May 1942.

MEAT.	TEA.	SALT	VEGETABLES	SHINTOGAR (SUGAR)	OIL
To 14th Novr 1942. Period 7 days.					
1910	7	45	5390	100	39.37
272.85	1	6.42	770	14.28	5.62
122782	450	2889	346500	6423	2529
126	.5	2.9	356	6.6	2.6
To 21st Novr 1942 Period 7 days.					
1280	11	54	6845	83	20
182.85	1.57	7.71	935	11.85	6.43
82282	706.5	3469	420750	5332.5	2889
84	.73	3.5	433	5.4	2.9
To 28th Novr 1942 Period 7 days					
1260	15	39	4785	74	46.12
180	2.14	5.57	683.57	10.57	6.58
81000	963	2506.5	307606	4767.07	2961
83	.99	2.58	317	4.91	3.05
To 5th Decr 1942 Period 7 days					
1460	15.5	41	4730	83	48.37
208.57	2.21	5.85	675.71	11.85	6.91
93856	994.5	2622.5	304069	5332.5	3109.5
97	1.03	2.73	316	5.55	195.75
To 12th Decr 1942. Period 7 days					
1590	4.5	181	4620	77	237.5
227.14	.64	25.85	660	11	27.96
102213	288	11622.5	297000	4950	12282

RICE.	MEAT.	TEA.	SALT.	VEGETABLES	SHINTOGAR (SUGAR)	OIL
To 17th October 1942. Period 8 days.						
6400	960	12.5	60	2460	71	28
800	120	1.56	7.5	307.5	8.87	3.5
360000	54000	702	3375	138375	3991.5	1575
507.62						
709	106	1.4	6.4	273	7.9	3
To 24th October 1942 Period 7 days.						
4560	590	11.5	45	1930	62	19.1
651.42	84.28	1.64	6.42	275.71	8.85	2.74
293139	37923	738	2889	124069	3982	1228
550.75						
533	69	1.3	5.2	225	7.2	2.2
To 31st October 1942. Period 7 days						
10080	1255	12	73	4100	81	36
1440	179.28	1.71	10.42	585.71	11.57	5.
645000	80676	769.50	4689	263569	5206.5	2315
871.28						
744	93	.8	5.4	303	5.9	2.
To 7th Novr 1942 Period 7 days						
9520	1600	12	40	3790	78	40
1360	228.57	1.85	5.71	541.27	11.14	5.
612000	102856	832	2569.5	243571	5013	260
970.42						
631	106	.8	2.6	251	5.2	2.

An example of records showing ration scale received by 'Anderson' Force in Hlepauk during October and November 1942.

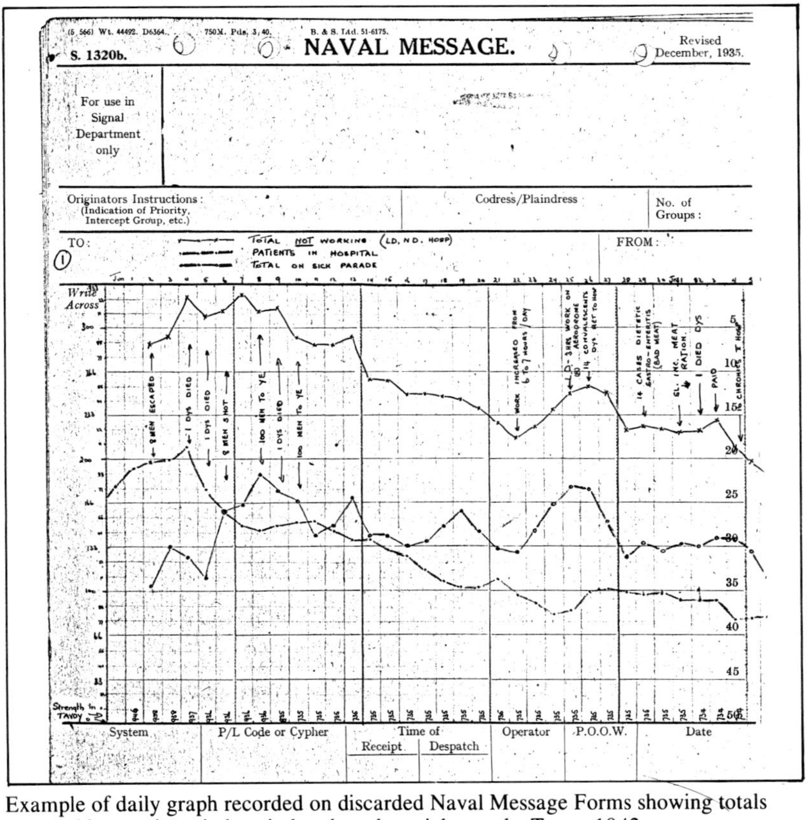

Example of daily graph recorded on discarded Naval Message Forms showing totals not working, patients in hospital and total on sick parade. Tavoy 1942.

Typical blood transfusion or saline giving set (left) used in POW camps. Australian War Memorial.

Elephant working with teak logs in Burma. U.S. National Archives.

Typical diesel engine and open trucks used during the construction of the Burma–Thailand railway. Post-war photograph taken in Kanchanaburi Memorial.

A group of 'Anderson' Force Medical Records kept by Jim Armstrong in Burma and Thailand.

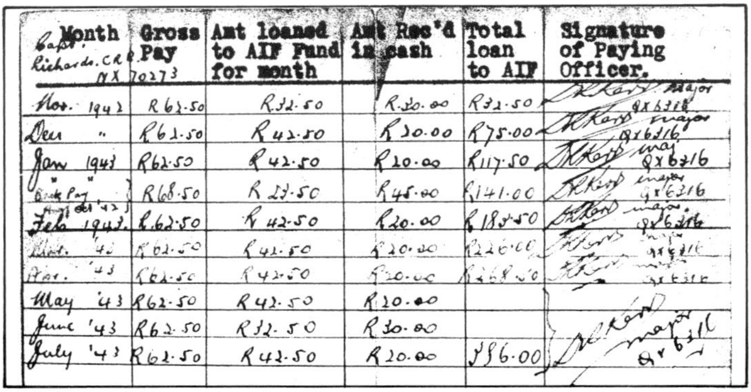

Month (Capt. Richards. C.R. NX 70273)	Gross Pay	Amt loaned to AIF Fund for month	Amt Rec'd in cash	Total loan to AIF	Signature of Paying Officer.
Nov. 1942	R62.50	R32.50	R30.00	R32.50	[illegible] major QX6316
Dec "	R62.50	R42.50	R20.00	R75.00	[illegible] major QX6316
Jan 1943	R62.50	R42.50	R20.00	R117.50	[illegible] maj QX6316
" Back Pay Aug/Oct '42	R68.50	R23.50	R45.00	R141.00	[illegible] major QX6316
Feb 1943.	R62.50	R42.50	R20.00	R183.50	[illegible] major QX6316
Mar. '43	R62.50	R42.50	R20.00	R226.00	[illegible]
Apr. '43	R62.50	R42.50	R20.00	R268.50	[illegible]
May '43	R62.50	R42.50	R20.00		[illegible] major QX6316
June '43	R62.50	R32.50	R30.00		
July '43	R62.50	R42.50	R20.00	386.00	

Pay book record of pay received after the Japanese deducted R60 for 'board and lodgings' and amount loaned to the Red Cross AIF Fund to purchase extra food for the sick.

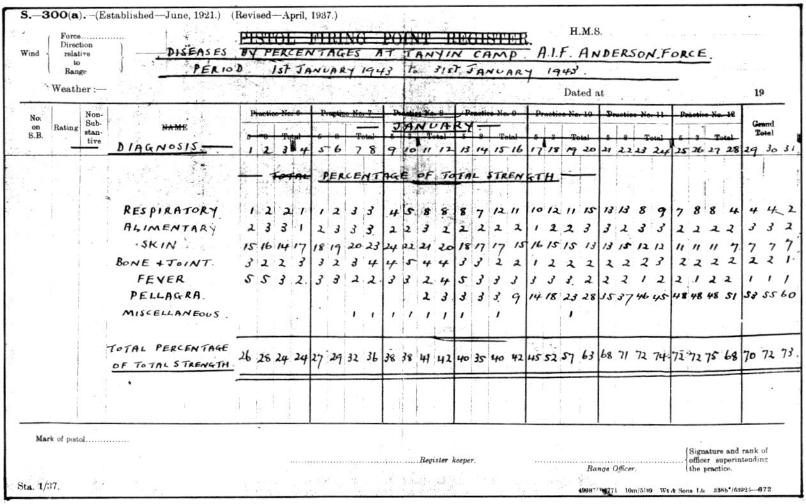

S.—300(a).—(Established—June, 1921.) (Revised—April, 1937.)

~~PISTOL FIRING POINT REGISTER.~~ H.M.S.

Wind { Force.............. Direction relative to Range }

DISEASES ~~BY~~ PERCENTAGES AT TANYIN CAMP. A.I.F. ANDERSON FORCE.

PERIOD. 1st JANUARY 1943 to 31st JANUARY 1943.

Weather:— Dated at 19

No. on S.B. | Rating | Non-Substantive | ~~NAME~~ | ~~Practice No. 6 … Practice No. 12~~ | Grand Total

JANUARY

— ~~TOTAL~~ PERCENTAGE OF TOTAL STRENGTH —

DIAGNOSIS	1	2	3	4	5	6	7	8	9	10	11	12	13	14	15	16
RESPIRATORY	1	2	2	1	1	2	3	3	4	5	8	8	8	7	12	11
ALIMENTARY	2	3	3	1	2	3	3	3	2	2	3	2	2	2	2	2
SKIN	15	16	14	17	18	19	20	23	24	22	21	20	18	17	17	15
BONE + JOINT	3	2	2	3	3	2	3	4	4	5	4	4	3	3	2	2
FEVER	5	5	3	2	3	3	2	2	3	3	2	4	5	3	3	3
PELLAGRA											2	3	3	3	3	9
MISCELLANEOUS							1	1	1	1	1	1	1		1	
TOTAL PERCENTAGE OF TOTAL STRENGTH	26	28	24	24	27	29	32	36	38	38	41	42	40	35	40	42

DIAGNOSIS	17	18	19	20	21	22	23	24	25	26	27	28	29	30	31
RESPIRATORY	10	12	11	15	13	13	8	9	7	8	8	4	4	4	2
ALIMENTARY	1	2	2	3	3	2	3	3	2	2	2	2	3	3	2
SKIN	16	15	15	13	13	15	12	12	11	11	11	7	7	7	7
BONE + JOINT	1	2	2	2	2	2	2	3	2	2	2	2	2	2	1
FEVER	3	3	3	2	2	2	1	2	2	1	2	2	1	1	1
PELLAGRA	14	18	23	28	35	37	46	45	48	48	48	51	53	55	60
MISCELLANEOUS			1												
TOTAL PERCENTAGE OF TOTAL STRENGTH	45	52	57	63	68	71	72	74	72	72	75	68	70	72	73

Mark of pistol..............

..............Register keeper. ....Range Officer. {Signature and rank of officer superintending the practice.

Sta. 1/37. 49987/9771 10m/5/39 Wt & Sons Ltd 338b*/63925—872

An example of the tables showing daily attendances on sick parade and the diagnoses (in percentage of camp strength).

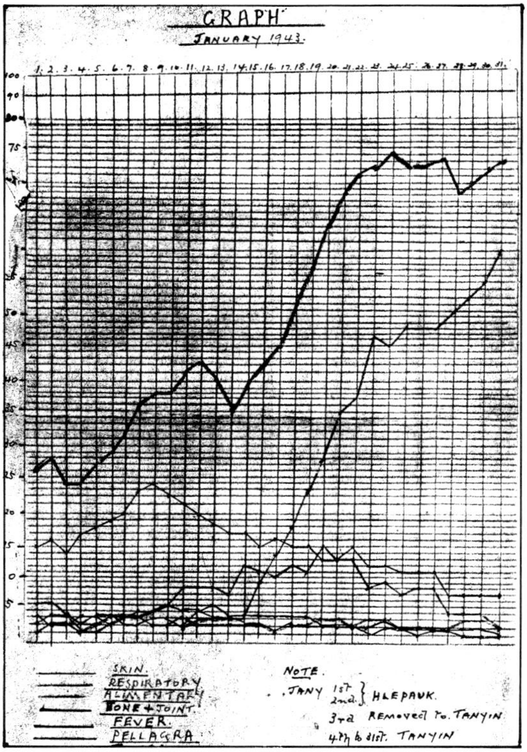

An example of the graphs showing daily attendances on sick parade and the diagnoses (in percentage of camp strength).

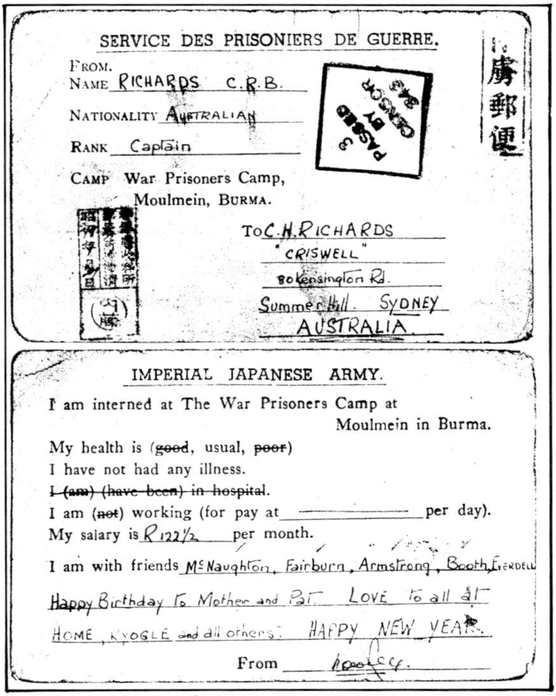

SERVICE DES PRISONIERS DE GUERRE.

FROM.
NAME RICHARDS C.R.B.
NATIONALITY AUSTRALIAN
RANK Captain
CAMP War Prisoners Camp,
Moulmein, BURMA.

PASSED BY CENSOR 3

俘郵便

TO C.H.RICHARDS
"CRISWELL"
80 Kensington Rd.
Summer Hill. SYDNEY
AUSTRALIA.

IMPERIAL JAPANESE ARMY.

I am interned at The War Prisoners Camp at Moulmein in Burma.
My health is (~~good~~, usual, ~~poor~~)
I have not had any illness.
~~I (am) (have been) in hospital.~~
I am (~~not~~) working (for pay at ______ per day).
My salary is R 122½ per month.
I am with friends McNaughton, Fairburn, Armstrong, Booth, EVERDELL
Happy Birthday to Mother and Pat. LOVE to all at HOME, KYOGLE and all others. HAPPY NEW YEAR.
From Rowley.

Message card written by Rowley on 27 December 1942 and received by his father in July 1944. The house was shown as 'Criswell' when, in fact, it was 'Cranmer'. Rowley's father recognised the message.

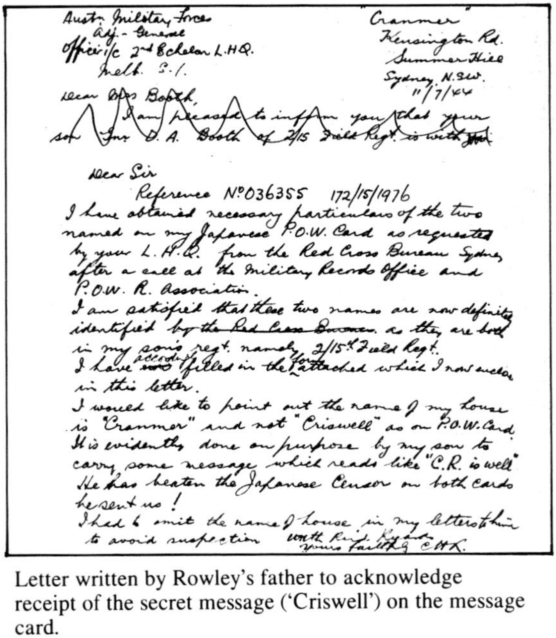

Aust. Military Forces
Adj.-General
Officer i/c 2nd Echelon L.H.Q.
Melb. C.1.

"Cranmer"
Kensington Rd.
Summer Hill
Sydney N.S.W.
11/7/44

~~Dear Mrs Booth,~~
~~I am pleased to inform you that your son Gnr D. A. Booth of 2/15 Field Regt. is with~~

Dear Sir
Reference Nº 036355 172/15/1976
I have obtained necessary particulars of the two named on my Japanese P.O.W. Card as requested by your L.H.Q. from the Red Cross Bureau Sydney after a call at the Military Records Office and P.O.W. R. Association.
I am satisfied that these two names are now definitely identified ~~by the Red Cross Bureau~~ as they are both in my son's regt. namely 2/15th Field Regt.
I have accordingly filled in the form attached which I now enclose in this letter.
I would like to point out the name of my house is "Cranmer" and not "Criswell" as on P.O.W. Card. It is evidently done on purpose by my son to carry some message which reads like "C.R. is well". He has beaten the Japanese Censor on both cards he sent us!
I had to omit the name of house in my letters to him to avoid suspection.
With Kind Regards
Yours faithfully C.H.R.

Letter written by Rowley's father to acknowledge receipt of the secret message ('Criswell') on the message card.

The bridge over the Menam Kwa Noi River, adjacent to the POW Camp at Tamarkan, being bombed by the U.S. Air Force. The bridge had been constructed by POW labour. The two arch spans were later replaced by box-girder spans. U.S. National Archives.

Pulau Damarlaut ('Jeep Island') as seen from the end of Jurong Road, Singapore. Site of the grave of Corporal S.R. Gorlick and the buried summary of Rowley's diary. Post-war photograph.

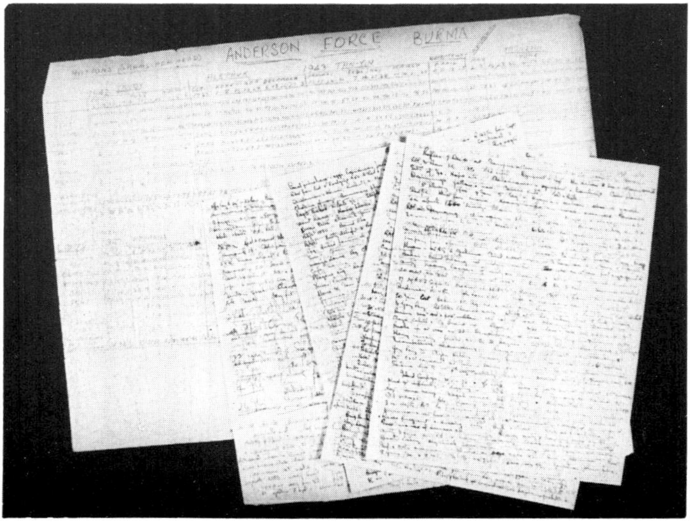

The summary of Rowley's diary and medical records buried on Pulau Damarlaut ('Jeep Island'), on 11 August 1944 and returned to Rowley by the Directorate of War Graves Services on 15 February 1947.

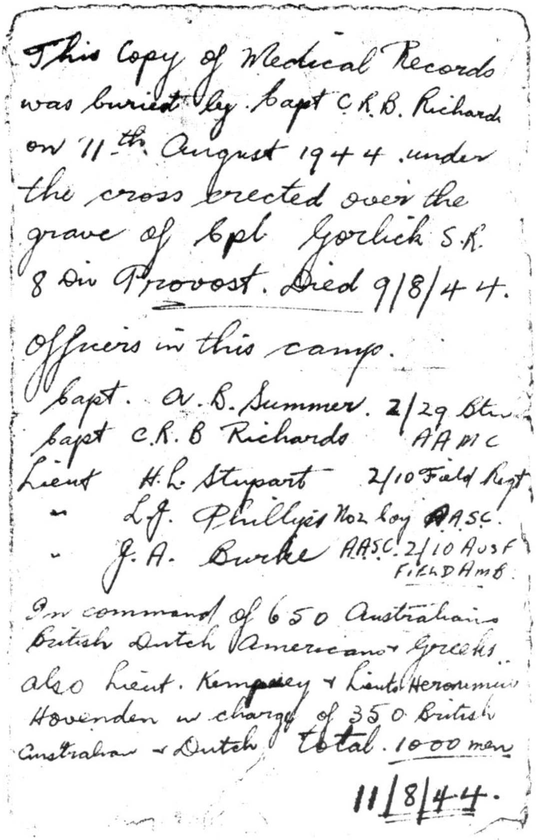

This Copy of Medical Records
was buried by Capt C.R.B. Richards
on 11th August 1944 under
the cross erected over the
grave of Cpl Gorlick S.R.
8 Div Provost. Died 9/8/44.

Officers in this camp.
Capt. A. B. Summer. 2/29 Btn
Capt C.R.B Richards AAMC
Lieut H.L. Stupart 2/10 Field Regt
" L.J. Phillips No2 Coy AASC.
" J.A. Burke AASC. 2/10 AUST FIELD AMB.

In command of 650 Australians
British Dutch Americans & Greeks
also Lieut. Kempsey & Lieuts Heronemus
Hovenden in charge of 350 British
Australian & Dutch total. 1000 men
11/8/44.

Message buried with the summary diary on Pulau Damarlaut ('Jeep Island'), on 11 August 1944 and returned to Rowley by the Directorate of War Graves Services on 15 February 1947.

Survivors from the sinking of the *Rakuyo Maru* being rescued by U.S. submarines. US National Archives.

Lieut Watanabe,
P.O.W. Camp Commandant,
SAKATA.

Sir:

The Medical Officers of the Sakata Prisoner of War Camp under your command wish to make the following report:-

1. The health of the men in this Camp is still unsatisfactory.

2. This fact is due to (a) Beri-Beri and Oedema.
(b) Diarrhoea.
(c) Dysentry still a cause for anxiety.

3. Beri-Beri and Oedema of which 27 new cases have appeared in the past four or five days, is increasing among the men at an alarming rate. The cause is a combination of Vitamin B1 deficiency and protein deficiency — both of which could be overcome by the suggested ration of fish or meat plus the full ration of soya bean and meso.

It is pointed out that the men have not received a regular meat or fish ration for more than two months.

It is noted that some of the earlier cases are improving with rest, Vitamin B1 injections and liquid.

4. Diarrhoea is also on the increase and has appeared during the past four or five days in combination with the Beri-Beri and Oedema. The cause is most probably the unbalanced ration (insufficient protein and fat for the amount of carbohydrate) and the rough nature of the green vegetables, (few leaves and many stalks).

5. There have already been two deaths from Amoebic Dysentry since arrival in Sakata, both of which could have been prevented by adequate emetine therapy. In the absence of emetine therapy five other cases now in hospital may also become acute and prove fatal. It is appreciated that every endeavour has been made locally to obtain emetine so we again request that endeavours be made to obtain supplies from the International Red Cross supplies in other P.O.W. camps in Japan. It is noted that the latest dysentry powders issued have had definite beneficial results on three of the four worst cases.

Signed. J.R. Roulston. R.A.M.C.
C.R.B. Richards. A.M.C.

30th October, 1944.

Letter to Lieutenant Watanabe, POW Camp Commandant, Sakata, dated 30 October 1944, drawing attention to the causes and management of beriberi, diarrhoea and dysentery.

Sakata POW Camp, Christmas 1944. Sleeping billets on left and hospital and RAP on right. British and Australian POWs assembled on the parade ground with Japanese guards and civilians. The worst winter in seventy years.

Australian survivors from the Sakata POW Camp on board the HMS *Formidable* returning to Australia, October 1945. *Back row (l to r)*: Stan Manning, Ern Wade, Jack Wade, 'Butch' Hall. *Front row*: Alick Robinson, H. Ramsey, Rowley, Keith Martin, 'Titch' Lemin, Kitch Loughnan. Royal Navy photograph.

Chukichi Takahashi (left) and Yuzo Matsumoto in Homer's Gardens, Sakata, 1958.

APPENDIX "D".

ISSUES OF CLOTHING & NECESSARIES WHILE AT SAKATA.

October 8th 1944	One large plate per man, one wooden spoon per man, six kettles. One tooth brush and powder per head.
November 19th 1944	One fundochi (loin cloth) one bar soap, one hand towel, 10 cigarettes per man.
November 6th 1944	One second hand shirt, one second hand working suit per man. (N.B. These were almost unserviceable).
November 24th 1944	One sheet, one pillow and case and one blanket per man.
December 15th 1944	One quilt and one blanket per man (2 extra blankets for men not receiving quilts).
December 16th 1944	One small plate, one metal spoon and fork.

N.B. Soap and latrine paper issued as mentioned in report.

An extract from the report on Sakata POW camp prepared after the cessation of hostilities on 15 August 1945.

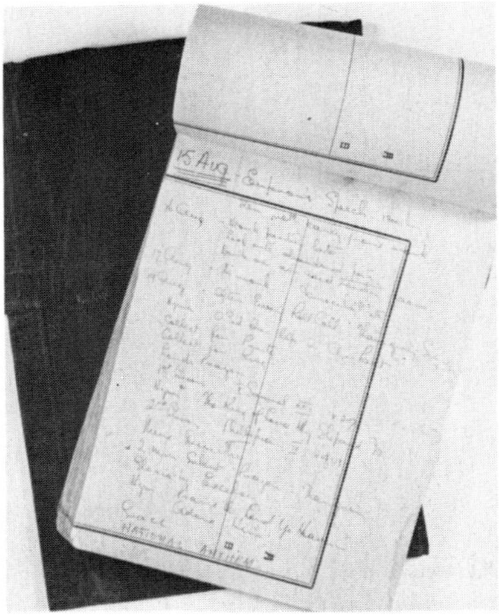

Extract of diary kept in an improvised note book in Sakata, Japan, from 15 August 1945. The improvised waterproof wallet was made in Tamarkan, as protection for the paybook.

Rowley, ex-POW, December 1945.

The HMS *Formidable*, carrying survivors from Japanese POW Camps, arriving in Sydney on 13 October 1945. The boom across the harbour was still in place. *Sydney Morning Herald*.

'Watch how Williams does it,' Drower advised. 'Waits until His Highness is jawing at someone else, then gives a quick flick of the wrist.'

'Neat manoeuvre,' Fitzsimmons agreed. 'I'd try it at home but wouldn't quite do for the ancestral rugs, old boy.'

'No rugs here, only bugs,' Rowley observed. 'Just imagine you're at the court of Henry the Eighth.'

'He beheaded people too,' Fitzsimmons growled.

The conversation had caught Naito's attention.

'What do you say there?' he demanded.

'I said "Good health to you",' Fitzsimmons told Naito, raising his glass.

The party lasted until eight o'clock when Naito finally slumped forward, arms and head on the table. Unfortunately it had not been possible to toss too much of the brandy under their chairs, so the younger officers were also fairly full. Rowley held no sick parade that night.

The smallpox began with sore throat, backache and headache followed by a red rash that was most profuse on the upper part of the face and scalp, the legs below the knee and the lower arms from elbow to hand. Then itchy and painful vesicles (blisters) began to appear, particularly on the scalp and the legs, from the third to fifth day. The temperature began to rise at the onset of symptoms and fever reached its peak at 105° F on the third day, dropped when the vesicles began appearing, then began to climb again, peaking on the tenth day when vesicles had developed into pustules.

There were five patients in the smallpox isolation hospital: three positive cases and two suspects. That was the height of the epidemic, a mild one as all troops had been vaccinated against smallpox, typhoid and cholera before leaving Australia.

Rowley, Jim, Pinky, Arthur (who had been released from line duties by the Nips), and Miles were up all night during the worst twenty-four hours when the patients were in a toxic state with the possibility of convulsions and secondary infection which could include pneumonia.

Apart from the constitutional disturbances, the fever and severe headache, the smallpox virus attacked the skin of its victims. At first the small septic spots were hard and about the size of a pea, but they became pus-filled blisters or pocks. The pustules dried, formed dark brown or black crusts, and these scabs, when shed, left permanent, pitted scarring of the skin.

'The virus is contained in the pustules,' Rowley told his orderlies. 'Infection is spread by droplets or by direct contact with the pus, and then when the scurf is being shed or from infected bedding and clothing. One of the Brits must have been in contact with a native with a mild dose. Or the contamination could have been on a shirt bought from one of the traders. The natives know enough about smallpox to burn the clothing of their dead but I wouldn't put it past some of the black-market hustlers to sell a dead man's shirt to one of our boys. We'd better warn the lads not to trade for clothing.

'Vaccination is good for three years, with some protection up to seven, but some of the Brits could have missed out on a booster.'

Jim McNiven had told Rowley that the Orientals discovered hundreds of years ago that smallpox didn't attack the same person twice, so they infected healthy people with a mild form of the disease so that they would be immune to severe attacks. A British woman traveller, Lady Mary Montague, introduced the idea into England. Later, of course, Jenner found that cowpox also gave immunity from smallpox and so became the father of vaccination.

'From *vacca,* Latin for cow . . . if that's of interest,' Rowley told his orderlies.

'Another interesting scrap of information from McNiven was that smallpox has something to do with the Japanese custom of bowing very low at first meeting. There was such a scourge of smallpox in medieval Japan that people's faces were terribly disfigured. On meeting an acquaintance who had been ill, or a stranger, one bowed very low, the next bows were progressively less deep so that one got a chance to get a quick peek at the other party's face. If the scars were very bad a polite person then continued the conversation with averted eyes so that they would not cause embarrassment by staring at the disfigurement or having to comment on it. Eventually it became part of ritual politeness not to stare another person in the face.'

'And with us, it's impolite, or being furtive, not to look a bloke in the eye when you talk to him,' Jim said thoughtfully. 'Yet we do our best to out stare the Nips all the time. That means we're being straight-out rude, doesn't it?'

'Sure thing. We're a lot of ill-mannered barbarians.'

'Well, so are they,' Pinky burst out. 'I'll have a good stare at every bloody guard on the line, and count his damned pock marks if I like. Only wish our staring hurt as much as their bashing.'

In the absence of disinfectant the sick were bathed with boiled water and made as comfortable as possible. Then the RAP staff scrubbed themselves down and dropped their shorts into a drum of water to be boiled over a camp fire outside the isolation tent. Rowley changed into his spare shorts and shirt and went to conduct the morning sick parade. Total sick excluding the smallpox cases was 100 out of a force of 500.

He was thinking about catching some sleep before the next parade when he saw Bill Drower walking away from the administration office. He was limping a little and Rowley called to him and waved for him to wait.

'Bill, you haven't seen me about that ulcer lately.'

'But it's doing fine, old chap. I'm keeping up the old treatment, you know. Not a thing to worry about.'

'So why are you limping? It's hurting, isn't it?'

'A mere trifle.'

'Damn your trifles. Get up on your cot and let me have a look.'

The small ulcer, no bigger than a threepenny bit, had formed a scab and should have been healing well, but Bill had obviously been on his feet too much. The periphery of new flesh was slightly inflamed and there was a yellowish trace of pus at the centre of the scab.'

'No Duties, one day, and you are to stay in bed,' Rowley decreed.

'But look here, Rowley . . .'

'No buts, Bill. I'm the doctor. You aren't going to move off your spine today. Pinky will bring your rations. Do I have to pull rank?'

'You're the doctor,' Bill agreed meekly. 'Truth is, it is a bit sore.'

About mid-morning Naito sent for the interpreter. Since the reason for the summons seemed trivial Bill wrote a short note of apology in which he asked if the messenger could do what was required. But Naito was in a bad-tempered stage of drunkenness. He sent for the day's work sheets and found that Bill was marked down as a worker.

Rowley had just dropped into exhausted sleep when Shimojo prodded him awake. He was wanted in Naito's office. When he arrived he found Colonel Anderson already on the carpet and Naito in a fury.

Naito accused Anderson and Rowley of falsifying the work sheets, marking sick men down as workers so that they would receive their pay. Rowley explained the situation quietly and carefully, hoping that it would sink in to the brandy-bemused brain. Naito calmed down, grunted and waved them to leave. It seemed that he was satisfied.

But the lull was neither before nor after the storm. It was in the middle of it. A little later Naito appeared in Bill's sleeping hut to inspect the ankle in question. He stared at the scabbed ulcer, backed off several feet in one move, opened and shut his mouth, and finally produced sound.

'Smallpox!' Naito screamed.

When Rowley tried to explain, once more, that the sore was a small tropical ulcer, he was told to shut up and the command was reinforced with the business end of Naito's revolver which wavered over the gathered company of Bill, Shimojo, Jim Armstrong, a couple of shrinking Korean guards and a lizard sitting on a bamboo beam. But the muzzle spent an uncomfortable amount of time pointing at Rowley's middle.

'If he isn't careful the damn thing will go off,' Rowley thought.

Naito, complete with armed guard, escorted Bill personally—at a safe distance—to the isolation hospital and saw to it that he was placed under the same mosquito net as the worst case, a young Britisher in the most unpleasant stage of the disease and covered from head to foot with stinking, scabby, suppurating sores. The patient had just been washed and the basin of washing water was standing near the entrance to the hut.

Having disposed of the patient, Naito insisted on being sprayed with disinfectant. Captain Hekking explained that he had none but Naito saw the basin of soupy turbid water by the door, mistook it for disinfectant, and insisted on being sprayed with it.

Hekking obeyed with commendable promptness, enthusiasm and vigour, making sure that the washings of the scabby sores entered Naito's nostrils, ears and mouth, as well as covering all exposed parts of his body. Naito departed and there was another lull.

Rowley arranged for Bill's bedding, clothes, eating tools and other odds and ends to be taken to him. Unluckily Naito appeared as the bedding was being carried over and ordered it to be carried to an open space in front of his office. Then he sent guards on the run to fetch Rowley and Colonel Anderson.

Naito had taken up his position in the middle of the office, booted legs straddled, revolver hanging by the trigger guard from his index finger, looking very much as if he was about to re-enact Custer's last stand. With much twirling of the revolver he told Anderson that he was a very bad man for falsifying work sheets and lectured Rowley on his horrendous behaviour in concealing such a frightful disease as smallpox and even refusing to admit that it was smallpox. He sentenced them to be stood at attention for two hours, with appropriate corporal punishment.

There was nothing to do but grin and bear it. The grin, often accompanied by a faint shrug, was common to most of the Australians in the camp now; a carefully contrived mixture of apparent good humour, with a slight curl of the upper lip that stopped just short of being a recognisable sneer. It was their only defence against the steady piling up of misery, of mud and toil, bashings and starvation. Men had learned to keep their hands at their sides when slapped or when watching a mate take a bashing, but they could grin. It disconcerted the Nips.

The guard who had been ordered to punish Anderson and Rowley did not seem to have his heart in his work. When he thought he heard Naito coming out of the office he administered a few slaps to the face or kicks to the shins. The rest of the time he contented himself with thumping his rifle butt on the ground and shouting, casting nasty aspersions on the pedigrees of the two offenders. Anderson stared straight ahead, grim-faced, tight-lipped. Rowley tried to maintain his grin. Standing at attention for two hours really wasn't so bad; the guards lining the route of a Royal procession could do it.

'If time isn't up soon I may have a slight accident,' Anderson murmured.

'So may I,' Rowley responded. 'But you haven't been on sick parade, Colonel.'

'Damn your sick parade. I don't need a doctor for a touch of colic.'

'I'll send Armstrong to your office with something. I can only supply one dose. Take the magnesium sulphate immediately and the Mist Astringens an hour later. After that try the powdered charcoal. We make our own supplies,' Rowley prescribed in a whisper.

'No talk!' screamed the guard.

Naito appeared and dismissed Anderson but he wasn't yet finished with the offending doctor. Rowley was ordered to personally supervise the burning of Bill's bed and all his effects. Pinky and Arthur brought kit and clothes from Bill's billet, they were piled on top of the bedding and dead grass was heaped all over the pile.

'We can't leave Bill naked. Despite Naito he's not headed for the cemetery,' Rowley said. 'We can get some of his things out while we're mucking around with this grass. I've already sneaked his diary and watch.'

Liberating some clothing was going to be more difficult as Naito had posted two guards. Unexpectedly help came from Naito's batman who nourished a healthy hatred of his officer. He came strolling from the side of the building and while pretending to inspect the preparations for the fire offered to remove Bill's clothing and blankets. What the two guards saw or thought was known to themselves alone. They stood at ease and chatted together, taking no notice of Naito's batman. Perhaps they thought he intended selling the articles; perhaps they shared his dislike of Naito.

The fire was just beginning to burn brightly when Naito weaved out of his office. He was drunk, but not too drunk to fail to notice that articles were missing from the bonfire pile.

'All blankets, clothing to be burned. Get them.'

'We did. They're on the fire,' Rowley told him.

Naito peered at the pile now being caught by tongues of flame.

'You tell lies again.'

'They don't seem to be there, do they?' Rowley agreed.

'Where then?'

'I don't know. I've been pulling dead grass to start the fire and it is a very dark night. Clothes gone . . . yes . . . but I have seen nothing,' Rowley said. 'Lieutenant Naito-san, best you ask your guards,' he suggested brightly.

Naito knew when he was trapped. He suggested that Rowley and his batman should make a search of the camp to find out where the missing articles had betaken themselves. They left at the double, made sure the search was extensive, and returned with an old pair of shorts, a couple of torn singlets and a blanket which was due to be destroyed in any case. Against their expectations, Naito had waited for them. He stood by and watched the burning until the last flames had died out, then ordered Rowley into his office again for half an hour of tongue bashing embellished with much brandishing of his revolver. Rowley was given to understand that his behaviour had been so bad that he should be placed in solitary confinement but Naito's regard for him as a doctor was so great that he would be punished with a mere beating.

One of the guards was told to administer the chastisement and Naito settled down with his bottle while the guard took Rowley outside. A young Korean who spoke some broken English, the guard was a quiet fellow who managed to convey politely that he was very sorry and the beating would be almost as painful to him as to the doctor-san.

Like hell it will, Rowley thought.

Then Yanagi, a much older Korean, materialised out of the darkness and waved the young guard aside.

Now, what? Rowley wondered.

Yanagi began shouting abuse and slapping his hands together very loudly just in front of Rowley's face, while stamping vigorously. After some minutes of this he stepped back, bowed politely, and jerked this thumb at the darkness. Rowley took off with more speed than dignity.

'Anderson' and 'Williams' Forces were moved from Kunknitkway to Anarkwan, the 45-kilo camp, and Naito provided the highlight of the move. After working right around the clock for three days the camp was ready to move. Train transport was to be provided this time as the line had been laid to Anarkwan but the train, which was due in the morning, did not arrive until late afternoon.

When everyone had clambered into the open steel trucks and the sick had been made reasonably comfortable, Naito arrived, so drunk that he could not walk unsupported.

He ordered everybody to detrain so that he, the representative of His Imperial Majesty, the Emperor, could board first. A clumsy 'grandfather' chair was carried aboard by the guards and Naito's batman lowered him into it. Then the first batch of prisoners was allowed to retrain. As the train pulled out Naito staggered to his feet and waved coyly to the assembled men waiting for the next trip. The wave was returned enthusiastically with a typical Australian gesture.

The surroundings of Anarkwan were beautiful: magnificent hills in the background on which morning and evening light made spectacular colour effects; the intervening dark green of jungle was lovely when seen from a distance, but a ghastly hothouse of decay to work in. A broad river flowed by the camp, shallow and clear with a bed of smoothed stones. The surface was thickly strewn with the purple petals of flowering trees and Jim Lynch was able to revel in his evening tong.

But the pressure was on. If the prisoners had thought they were under pressure beforehand, they began to learn what Japanese slave-driving could be.

The word was that Thanbyuzayat was being attacked and a stick of bombs had just missed a fuel dump in the jungle. The Japanese were suffering reverses on the sea, in the air and on land, according to the secret dicky bird. The Australians were more than holding their own in New Guinea. The Japs had evacuated Guadalcanal and in the battle of the Bismarck Sea in March the Japanese invasion convoy had been destroyed.

The Japanese High Command had decreed that the Burma-Thailand railway must be finished quickly. Work started at 0830 hours and did not finish until 2230 hours. Line was being laid at one to one and a half kilometres a day. The Storm Trooper was in charge of the sleeper carriers and looked, Jim Lynch observed, like a poor relation of a Mongolian raider.

Jim was with a spiking gang that was by now expert at swinging eight-pound hammers and their guards, seeing that they knew what they were doing, did not bother them too much. But there was not time for the nightly tong. The men collapsed into sleep with the mud of the embankments caked on their legs. Not having time to bathe and shave upset Jim.

Naito held his own sick parades: test for eyesight, a slap in the face; bad legs, a push; diarrhoea, a poke in the abdomen. Rowley argued the case for every man. When Naito demanded 700 men for the working parties and only 500 could be supplied he had Colonel Anderson locked up and threatened to shoot him. The threat was not carried out and the Colonel was freed as Naito knew well he was needed for camp administration. The work quota for the line layers was increased to thirty-six trucks a day. And rations decreased as the work increased.

Non-workers were to receive neither pay nor rations. So 700 men had to be sent out, the most fit almost carrying sick comrades so that they could at least appear on the line and draw pay. There was no way that Don Murchison's locally contributed Red Cross fund could buy enough food for all the sick.

For refusing to send sick men to work, John Williams was stood outside the guardhouse, bare-headed in the tropic sun, for twenty-four hours. Except for one occasion of about forty-five minutes when he collapsed and had to be revived, he remained at attention.

The Force was moved from Anarkwan to the 60-kilo camp, Taunzan. It had been a native coolie camp and not only was it horrifyingly filthy, but there were still dying natives in the huts, dead bodies lying in the fringes of the jungle, and a cemetery of over 200 graves within a few yards. The graves were shallow and there was the stench of death. Sergeant Shimojo, who was in charge of the first parties to reach Taunzan, liked the situation as little as John Williams did. He would have moved on if there had been any place to which the Force could be taken, but such a large troop of men could not just move into the jungle. Williams set about cleaning up, organising burial details and establishing sanitary facilities. The Burmese had just squatted wherever the call of nature came to them.

Naito was ordered to Thanbyuzayat to attend some Japanese ceremony and did not return. It was said that he had disgraced himself by collapsing drunk at the ceremony and Nagatomo had sent him back to Japan for punishment.

'Anderson' and 'Williams' Forces did not mourn his loss.

Jim Lynch, Scippy and Rusty Irwin were packing soil under sleepers; back-breaking work, wielding fifteen-pound hammers. Then they were moved to a bridge-building detail, manhandling heavy timber in the wet and hand-driving piles into a creek. Jim's bouts of malaria were more frequent and his ribs more prominent but orders were that men previously classified as Light Duties had to work.

The Dodger, who was a chronic unfit, with recurrences of the pellagra rash, managed to get himself assigned to assisting the cook in the Nip kitchen where he sometimes received 'presentos' of extra food or stole it. The great worry of his existence was that the disease might render him incapable. Fancy surviving the bloody line then finding he couldn't have a bit when he got back to civilisation. Not surviving the bloody line didn't enter his head.

Roy Whitecross was on No Duties. Although the tropical ulcer had healed, he had developed his first attack of dysentery which lasted five days and left him so weak that he could not walk twenty yards. John Shaw was suffering from intermittent bouts of diarrhoea and losing condition. Rowley also had diarrhoea and admitted to himself that he felt lousy.

The news now was that 7,000 AIF from Changi were working towards them, between the 113- and 130-kilo camps on the Thai side of the border, part of a combined-nationalities force of POWs who had been working on line-building from Bangkok. Fifty thousand of them altogether, according to the rumours. There was also said to be considerable daily troop movement by the Japanese from Bangkok to Moulmein and to and from Amherst on the coast. Were they expecting an Allied landing?

Rowley and Charles Anderson clashed again over the sick as the attacks of malaria became more severe. Anderson did not or would not accept the significance of cerebral malaria. 'Everyone in the tropics gets malaria. It's nothing serious,' he said. 'Mosquitoes have to live too, you know, Richards.'

As the position became worse he began to recognise the existence of sickness but insisted there was nothing he could do in the face of the Nip demands, except send the sick out. If they didn't work they would starve. The dilemma was daunting.

'I've got to talk the Nips into understanding . . . or let them see for themselves, when men drop on the line, that we are a spent Force and must be given a month's rest. We must start making representations to Hamilton at Thanbyuzayat.'

On 27 May 1943 Rowley began writing a report to Colonel Hamilton, the senior medical officer at headquarters in Thanbyuzayat. After describing conditions at Kunknitkway and Anarkwan he continued:

> This camp, Taunzan, on our arrival, was being evacuated by Burmese. The filth was appalling and the flies countless. About 100 yards from the camp is a cemetery, in which there are about 150 graves which, from the first few days, were increased daily by 2 or 3.
>
> The cause of death was originally reported by the Japanese as Cholera and Smallpox; I have considered the possibility of Malaria and/or Malnutrition as contributing factors.
>
> Within the first few days of our arrival the incidence of Diarrhoea increased considerably. A noticeable development at this stage was the vomiting which accompanied the Diarrhoea. This suggested the possibility of some irritant in the food. Mouldy rice, which we have had till a few days ago, and green egg fruit, which were eliminated from our diet, were held suspect . . .
>
> After 10 days in the camp, cases of Primary Malaria commenced to appear. These men usually presented with severe pains behind the eyes and pains in the knees, hips, back and other joints. These symptoms were accompanied by fever (101 to 105 degrees), sweating, shivering and in most cases by Diarrhoea. Most cases responded to quinine within 36 to 48 hours.
>
> During the past eight days there have been some 80 cases of Primary Malaria. With the arrival of new supplies of quinine 3 days ago all men are receiving 8 instead of 4 grains daily. Dr Higuchi, who visited us yesterday, ordered us to have 8 and 4 grains on alternate days.
>
> Yesterday 49 sick men from 'Anderson' Force and 50 from 'Williams' Force were sent to work and were employed on the same task as the others and for the same duration.
>
> Four of 'Anderson' Force were returned by the Japanese, and some 12 to 15 were allowed to rest during the afternoon. On returning to camp some 20 were in a state of near collapse.

Jim Armstrong stuck his silvery head through the doorway of Rowley's billet.

'Gunner Dare isn't too good. I think you'd better come.'

It was 0900 hours.

Dare lay on his side on his section of the bamboo platform in the hospital hut. His knees were drawn up in pain against his abdomen and he was shivering so much that his teeth chattered. He moaned weakly. He showed symptoms of dehydration, his skin was devitalised and a ghastly yellow-grey colour.

Rowley straightened out his legs and palpated the abdomen which was tender. He ascertained that Dare had had his appendix removed in childhood. Rowley began massaging one of the cramped legs with long, sweeping upstrokes and nodded to Jim Armstrong to get to work on the other leg.

'Fill all the hot water-bottles we have and pack them around him,' he told Pinky.

As he worked Rowley reviewed the case. He had seen Dare on sick parade the previous evening at 2100 hours after the lad had come in from the line, and had administered 1 cc of Pestis vaccine recently supplied by the Japanese to combat the possible occurrence of plague. Dare had said he felt tired and had a bit of diarrhoea, about four to six rather watery daily motions for the past three days. Otherwise he seemed to be in fairly normal health and said he felt he could go on working. There were other chaps who had had to rest during the afternoon and needed a day off more than he did.

At midnight a mate had brought Dare to the RAP. He felt ill and complained of abdominal pain. He said he had had four small, watery motions since 2100 hours and had vomited a little but thought he was in for an attack of malaria until the bad abdominal pain started.

Rowley had admitted him to the hospital hut and given him kaolin and charcoal. Pinky Rhodes was to stay with him through the night. At 0400 hours Pinky had awakened his RMO. The pain had not been relieved and Dare was distressed. His condition warranted an injection of one quarter grain of morphia from the RAP treasure chest. He had slept and at 0800 hours Pinky reported that Dare was resting quietly, had had no further motions and had not vomited.

Now cramps, chills and severe prostration. If this is malaria I haven't seen anything like it, Rowley thought. The heat and massage seemed to be having an effect. Dare's muscles relaxed and the shivering stopped. Rowley remained with him for an hour and his condition seemed stable.

At 1100 hours Sergeant Sowten, 2/10 Field Regiment, reported to the RAP with abdominal pains and vomiting. He had a history of several days diarrhoea and was on No Duties. Rowley took him immediately to the hospital.

'How's Dare?' he asked Jim.

'I was just going to call you. The pain must be starting again. He's getting restless and drawing up his legs. But he's smiling. Damndest thing!'

'Smiling?'

'Yes, grinning as if he were really happy about something. Just started.'

Dare's flesh was cold again and markedly yellowish-grey. When pinched between the thumb and forefinger the skin remained puckered. But he was indeed smiling. The facial muscles had gone into minor spasm producing a mild *risus sardonicus*. It was disconcerting. The lad was suffering pain yet lay there smiling up at his attendants.

At midday Dare's condition deteriorated suddenly and he died in twenty minutes.

The body had barely been carried out to the mortuary hut when Jim called, 'Sowten's going the same way. Cramps and chill.'

'Get going with the massage and hot bottles and get some fluid into him,' Rowley ordered. 'Anything the kitchen can supply, boiled water, liquid from the stew. I'm going to try an intravenous infusion but first I have to see Anderson and Shimojo and the doctor at the Nip camp. I think this is cholera.'

Abdominal pains, cramps, chilling, rapid dehydration and devitalisation (although the body was not apparently losing fluids), sudden collapse and death; the symptoms seemed all wrong. Cholera was caused by the introduction into the alimentary canal, most usually by way of contaminated water, of millions of *cholera vibrio*, a curved bacterium that irritated the stomach and bowel causing continual vomiting and diarrhoea, with copious 'rice-water' motions as often as every ten minutes, until the body had cleared itself of the bacteria or, as happened if the disease was untreated, the victim died of dehydration. But there had been almost no vomiting or diarrhoea. The onset of the illness and its progress to death had been more rapid than anything Rowley's reading or brief training in tropical medicine had prepared him to expect. Yet he believed that it had to be cholera.

An engineer had died in the Japanese camp three days before Dare became ill and it was said that a doctor had arrived there from Thanbyuzayat to make tests. The doctor turned out to be a laboratory technician who was taking rectal smears, and he confirmed Rowley's diagnosis. It was cholera but of a particularly virulent form, *cholera sicca*, or dry cholera, that created a toxin so rapidly that the alimentary canal was paralysed and the body could not rid itself of the *cholera vibrio*. Hence the rapid deterioration to death.

The Japanese were terrified of cholera. For once there was no lack of cooperation, no bargaining, no argument: they were willing to do almost anything Rowley asked.

Rowley appropriated a disused buffalo stable, nothing but a roofed structure with railed stalls but no walls, and Sergeant Wilstencroft and his staff began the job of fitting it out as an isolation hospital with walls, fireplaces, latrines and a series of raised decks, each carrying four to six patients, at intervals of about nine to ten feet. These were so arranged that the patients with cholera, dysentery and severe diarrhoea could be isolated in small groups until they were found to be positive bacteriologically and clinically.

During the afternoon of 27 May Gunner Childs of 'Anderson' Force developed symptoms of *cholera sicca* and by evening another man, Private Lasham, was ill. Sowten and Childs died on the following day and Lasham a day later, all smiling the disturbing smile of *risus sardonicus*.

Sardonic (Latin root: *Sardonius*—scorn) was an adjective first used by Homer to describe bitter, scornful, mocking smiles or laughter. The word was derived from *sardopin*, a fabled herb the eating of which was said to produce facial convulsions resembling horrible laughter, followed by death. *Risus sardonicus* was an involuntary spasm of the facial muscles, yet Rowley could not get the uncomfortable fancy out of his mind that it was the final manifestation of the tiny, scornful smile that the prisoners, tortured by exhaustion, disease and brutality, had developed as their only defence against their tormentors. Was it all they had left: a scornful smile for the Japanese and for death?

On 28 May, after Sowten died from *cholera sicca*, the Japanese lab technician took rectal smears from the whole camp and prepared single-plate cultures for all hospital patients and suspects and cultures in groups of five for the remainder of the camp.

By the following day all suspects and positive cases had been moved into the isolation hospital. The volunteer isolation hospital orderlies were confined to the hospital with their patients for the duration of the epidemic. They were: Jim Armstrong, Pinky Rhodes, Arthur Baker, Jackie McKone, Harry Moxham, Bob Kemp, Woodbridge, Keith Fisher, Johnson, Arthur Harris, Don Booth, Duncan McNaughton, Don Bertram, Allen Taylor and Jamieson of 'Anderson' Force; Kellie, Kirby, Waters and Miles of 'Williams' Force. McNaughton, Rowley remembered, was from Kyogle.

The isolation area was strictly forbidden to all except medical staff or anybody authorised by a medical officer. As Colonel Eadie was ill Rowley was the only medical officer available to handle the epidemic. There was no magnesium sulphate available or procurable, no disinfectant, no distilled water for the preparation of infusions, no infusion equipment, and Rowley was in the heartbreaking situation of practising medicine without medicine or equipment.

Private Critchley of 'Williams' Force developed the normal cholera symptoms.

'He's got to have fluid,' Rowley told Sergeant Wilstencroft. 'I can make up a saline solution but I need equipment for the infusion—a bottle, tubing cannula—a hollow needle for insertion into the vein. What can you do?'

Wilstencroft acquired a beer bottle, per favour of the Dodger. Rubber tubing was cannibalised from a stethoscope. The needle Wilstencroft made from a piece of copper piping supplied by the driver of a Jap truck.

Rowley attempted the infusion of saline solution on Critchley in a cloud of flying ants which were then plaguing the camp. The bore of the needle was too small and he was able to give only 30 cc intravenously. Someone suggested making a needle from fine bore bamboo which could be sharpened to a hard, sharp point. Wilstencroft began experimenting, making wider bore needles of copper and of bamboo.

'In the meantime keep getting fluid into Critchley,' Rowley told Jim. 'Make him drink . . . any fluid the kitchen can supply.'

Jim was scowling. He rubbed a hand over his deeply creased forehead.

'What's the matter? Tired . . . or what?' Rowley asked.

'I've got a headache. It's a bit fierce.'

'Diarrhoea?'

'A bit more than usual. Pinky's been off colour for a couple of days, too.'

Now that he had the opportunity to think about it, Rowley realised that he had a headache also.

'Let's hope our innoculations are proof against this thing, Jim. We have to keep going.'

Wilstencroft produced several usable needles and Rowley was able to give Critchley a further infusion. The lad was battling on.

The culture plates were returned by the Jap technician. Twenty-five groups, representing 116 men, had positive cultures. All suspects were isolated and individual tests made.

Roy Whitecross and one of his mates were among the suspects. During the first test there had been five samples to a slide. Roy and his mate, Jim, were both in the same group.

'Which means either of us has one chance in five,' Roy told Jim. 'This waiting for the individual results isn't too pleasant. Anyway, I reckon, if we had it, we'd know by now.'

Roy was determined to believe that it couldn't happen but he was feeling sick, worse than he had ever been. He felt nauseated and couldn't manage to eat. He'd had diarrhoea badly since the Force moved to Taunzan, thought he had lost about a stone and a half in weight, and was looking a bit of a wreck. Well, he could only wait and see.

By 2 June there were thirty-five patients in the isolation hospital of whom sixteen were bacteriologically or clinically positive cases of cholera. The worst cases were on drips, equipment manufactured by Wilstencroft. Critchley and Chamberlain, both serious cases, seemed to be improving.

Cholera was transmitted through the faeces of carriers and could be borne by flies, passed by dirty fingers, or carried in contaminated river water. Among the Burmese there was a reservoir of cholera to which they had built up a certain immunity but which flared into periodic, vicious epidemics. There was no doubt now what had killed over 200 Burmese in filthy Taunzan.

Rowley had written a list of observations and recommendations to Colonel Anderson on 31 May:

> All existing latrines to be closed and new ones constructed immediately. The ex-Burmese latrine will take preference . . .
>
> The isolation hospital is surrounded by masses of decaying rubbish. The ground surrounding this hospital must be bared of all rubbish and greenery . . . the cattle living at the end of the hut must be removed . . . a water point with chlorinated water must be established at the entrance to the Suspect Cholera Isolation Area . . . food must be taken over at the entrance to this area.

There were further recommendations for cleaning up around the billets, the closing of existing rubbish pits and building of new ones and the maintenance and deepening of drains. John Williams supervised the sanitation improvements.

All food was cooked in the camp kitchens but separate utensils were used for the isolation hospital. The kitchen staff carried cans of food to the barrier built where the water point had been established, poured the contents into the isolation cans and retired. The isolation orderlies then took their cans. Rowley, officially, was the only person who was permitted to move between the isolation area and the main camp, but there was one unofficial visitor whom he did not see and decided he did not wish to see.

As he was scrubbing at the chlorinated water point after checking the drips he noticed a tremor of the long grass beyond the barrier. 'Snake in the grass,' Rowley said loudly.

Pessimistic Pete Smith, the Roman Catholic padre, lay flat on his belly in the grass and held his breath. He was blaming himself vigorously for timing his visit so badly. Rowley was an alert young fellow, generally even-tempered but liable to do his lolly a bit over contraventions of medical orders. Perhaps if he hugged dirt for a while Rowley would really believe that he had only seen the ripple of a snake passing through the grass.

Rowley knew all about Padre Pete. Jim had told him who was coming in to wash and shave the sick.

Colonel Eadie, the medical officer attached to 'Williams' Force—an elderly ear, nose and throat surgeon who should have been kept at the base hospital where his specialist skills were very much needed—was doing his best to cope with the RAP and sick parades for the combined forces while Rowley handled the cholera epidemic. But he was struggling against an outbreak of malaria in a very severe form. Private Thompson of 'Williams' Force had had the fever for two weeks and was not responding although hospitalised and on quinine. He was bleeding from the nose, and was a lemon-yellow colour. He was unconscious for twenty-four hours, then quietly died.

Private Haines, also of 'Williams' Force, was a similar case. He had worked until he was completely exhausted and had to be persuaded to report on sick parade. He had thought he was not sick enought to occupy time while others were much sicker, but for four days he was difficult to feed and was gradually becoming more dehydrated. His body temperature soared suddenly to extreme fever (hyperpyrexia), and sweating profusely, he quietly died.

Jim Lynch's individual cholera test had proved negative but he was also hospitalised with the most severe attack of malaria of his history. He complained not at all and was the most cheerful patient in the hospital. Still, Sergeants Charlie Clayton, Arnold Noble and Scippy Maher worried for him and Noble caught Rowley between the isolation hospital and the RAP.

'Doc, you have to write lots of reports. I've been thinking you must be due for a new pen. I don't write much, with a couple of printed letter cards so far being the Jap form.'

He produced his fountain pen.

'It's nice. How much?' Rowley asked.

'Seven and a half rupees would about do.'

'Sold.'

As Rowley guessed, the money went into the communal fund and was used to buy extra food for Jim.

Two malaria cases were suffering from delusions and in their delirium had to be constantly watched. One managed to get out of the hospital and out of the camp but was caught before he had wandered too far into the jungle. Sergeant Shimojo was persuaded that he was mad rather than an escapee so he was returned to the hospital instead of being shot, but he died, burned up by hyperpyrexia. Colonel Eadie, in ill-health for most of his time in the jungle, was worrying himself to the verge of collapse about the disease which was right outside his experience.

'Stop worrying yourself sick, Colonel. We can't do without you,' Rowley told him. 'We have cerebral malaria now and I'm scared stiff that this is about the right time for pellagral dementia to start showing itself. We'd all better start praying.'

Sergeant Westgarth of 'Williams' Force had been sick for three days before he was admitted to the isolation hospital as a cholera suspect. His individual cholera test was positive and his condition was serious, with vomiting and copious rice-water motions. Rowley began administering coramine (a heart stimulant) from his own personal kit four hourly, kaolin quarter to half hourly. The orderlies were administering fluid by mouth, soup and even hot chocolate. He appeared to be improving, then there was a sudden deterioration in his condition.

'We'll give him saline solution intravenously, 300 ccs,' Rowley decided.

The apparatus was a bottle with the bottom knocked off, a 10-cc syringe barrel as a drip control, stethoscope tubing and one of Wilstencroft's needles, made from fine brass tubing, as a cannula. Rowley had a headache and did not immediately notice, as the equipment was set up, that there was an unauthorised person in the isolation hospital. When he did notice he snapped, 'What the devil are you doing here?' at Colonel Williams.

'Unofficial volunteer orderly,' John Williams told him.

The cannula was inserted and the drip functioning when Rowley felt a wave of dizziness. The hut was revolving. 'Keep that bloody needle in the vein,' he sang out to Jim Armstrong and fainted.

John Williams bent over the RMO's body sprawled on the dirt floor. 'Do you think he has cholera?' he asked.

'Yes,' Armstrong said, never taking his eyes off the steadily held cannula. 'The doc has cholera. We all have.'

9

Things Couldn't Get Worse

The *cholera vibrio*, swift and vicious in its attack, was also quick to die out. Of the twenty-five positive group plates representing 125 tests, there were only seven positive single plates. Of the sixteen cholera cases placed in the isolation hospital, seven died.

In retrospect there was little doubt that not only Rowley and his orderlies, but the greater proportion of 'Anderson' and 'Williams' Forces at Taunzan had suffered mild, abortive attacks of cholera during early June 1943. The immunity built up by innoculation, combined with rigorous attention to hygiene and sanitation, had saved them from the devastation caused in other camps on both the Burmese and Thai sections of the line.

Rowley received, by the hand of a driver, a letter from Adjutant Stuart Ward at one of the forward camps with news of fourteen cholera deaths in a camp of 1,900 men, as well as many severe ulcer cases and an almost 100 per cent incidence of malaria. Colonel Anderson's correspondence brought him news that was even more grim. Captain Lloyd Cahill, the RMO of 'F' Force on the Thai section of the railway, wrote that his group had ninety-seven cholera deaths.

Truck drivers and working parties passing through Taunzan on their way to other locations told of camps where the cholera had been so severe that it had become impossible to bury the dead. In Burmese and some Dutch and British camps the bodies had had to be piled up and burned.

The remaining fit of 'Williams' Force were sent to work at the 40-kilo camp on 7 June, leaving Rowley a legacy of their fifty sickest patients. Altogether, in the isolation hospital and main hospital hut he had ninety-three seriously ill men.

Don Murchison had been able to send some Red Cross funds. (from the POW local fund, since no funds from the International Red Cross had been able to reach them): the funds to be used, on Colonel Hamilton's instructions, for the purchase of food for the sick. The entire amount was R 258.80 which purchased 118 dozen eggs and a small quantity of sugar, fruit and tinned milk, as well as a few fresh chickens. With these supplies and the ox ration the cooks, despite makeshift ovens and cooking utensils, were able to produce convalescent diets: chicken broth, ox soup, egg custards flavoured with syrup made by scraping the pulp from the skins of pineapples and boiling it with sugar.

Corporal Jamieson, 8 Division Signals, who was unable to work on the line because of a mild degree of corneal blindness caused by beriberi but was otherwise comparatively fit, acted as Rowley's caterer. Together they made the hospital rounds each morning, Rowley nominating the diet for each patient and Jamieson making a list, in the absence of paper, on a piece of bamboo which could be rubbed down and reused each day. Between the difficulty of making clear markings on the bamboo and his defective vision Jamieson often got into difficulties reading his own writing. But he always succeeded in acquiring the necessary items from the canteen, supervising the preparation of the invalid food and producing the prescribed diet for each man.

Despite the passing of the cholera epidemic and the improvement in diet the troubles at Taunzan were not over. Rowley had felt things could not get worse but was beginning to realise that that was an overoptimistic outlook.

The deaths of Privates Thompson and Haines were only the beginning of the malaria epidemic. Many of the recurrent malaria cases, Rowley noted, showed herpes around the mouth and four or five new cases were showing mild pellagral skin and tongue signs. Indications were that the camp would be hit not only by malaria but by pellagra in the second, or killer, stage.

Although Rowley had recovered from his fainting spell in the isolation hospital within minutes and had even been able to gather himself up off the floor and finish the job of Westgarth's infusion, he was not feeling fit, which irritated him. He stubbornly refused to admit weakness and exhaustion in himself although he watched his men with the anxiety of a hen with chicks. On the eve of his twenty-seventh birthday he, John Shaw and Jim Armstrong had finished off the last of his flask of brandy and on his birthday the kitchen and RAP staff had presented him with a cake concocted from ground rice and chindegar, with icing made from chindegar and ground rice, even spelling out 'Happy Birthday 27'. It had been a morale-booster but he could not entirely ignore the continuing headaches combined now with backache and a slight fever. He had to admit that he probably had malaria and certainly felt lousy.

Daily more and more malaria cases required courses of quinine. Over two weeks the courses had to be reduced from 32 grains for ten days to 32 grains for one day, followed by 24 grains for three days and thereafter 16 grains for three days. The next reduction was to 24 grains for one day followed by 16 grains for four to six days. This drastically reduced dosage was inadequate, and faced with a complete cut-out of supplies before the end of the month, Rowley was making urgent representations for more quinine.

Private Rodd of 'Williams' Force was so ill with malaria on 17 June that it did not seem that he would last the night. Jim Armstrong stayed up with him so that Rowley, in the grip of malaria himself, could sleep a little. He greeted Rowley the following morning with: 'Rodd is still with us'. It was one of the most astounding examples of stamina and the will to battle on that Rowley had yet seen. But Rodd's great battle ended by hyperpyrexia and sweating thirty-six hours later. Rowley's provisional diagnosis was exhaustion, malnutrition, dysentery and malaria, an almost undefeatable combination that would reap a considerable harvest before the Wet season was over.

Five days after Rodd's death another malaria victim, Private Marven, Australian Army Service Corps, died after being ill for three days. His complications were chronic amoebic dysentery and exhaustion.

The sole advantage of the Wet was that the rivers were all in spate, making bridge construction impossible, so work on the railway had been temporarily suspended and the few fit (on one day only seventy of the entire force of 1,000 were able to work), were employed on repairing the roads which were not much more than tracks. A disadvantage was that it was also impossible for supplies of food to be sent through to the forward construction camps. Had it not been that Taunzan was only a couple of kilometres from a food dump and 'Anderson' Force were given extra rations of perishables which could not be sent further along the line, they would have been starving.

Those who could not receive the food probably were starving.

The first week of June had seen the worst of the cholera epidemic; the first week of July saw the misery of the malaria epidemic with 309 out of 405 sick suffering malaria with recurrences every two or three weeks. Yet the Nips were desperate to make up lost work time and again blitzed the sick parades, demanding 300 workers, at least, out of the combined forces. It was only with difficulty that 200 sufficiently fit were mustered from 1,000 men.

On 8 July Signaller Beckford, 8 Division Signals, died from malaria, dysentery and malnutrition. Gunner Key, 2/10 Field Regiment, the last case of cholera to develop, in late June, had received infusions of saline solution and had seemed to be in a satisfactory condition. But he suddenly gave up the unequal battle early on the morning of 10 July: dysentery and pellagra, as well as cholera, had been too much. Beriberi oedema increased daily and jaundice was almost universal among those suffering from recurrent attacks of malaria. Rowley was in bed for several days, but when Jim took ill for two days he returned to duty despite jaundice. At least he was not personally worried by the food shortage. He found it difficult to eat or even to bear the smell of food near him.

Two cases of second-stage pellagra were suffering from dementia, one with a persecution complex that he was to be shot for trading with the Burmese and the other with delusions about time, place and position. It seemed that an epidemic of pellagral dementia was to be added to the malaria.

'What will happen to our sick?' was Rowley's sole agonising thought when instructions were relayed from the Japanese engineers that the Australian force at Taunzan was to move to the 70-kilo camp, Mizale.

Moving-day was sheer torment. Although the railway was only 500 yards from the camp the track was inches deep in mud and all gear, including the heavy kitchen equipment, had to be carried to the entraining point. The fit were to go to Mizale. The sick were to be sent back five kilometres to the 55-kilo camp where a permanent line hospital had been established.

Forty-one stretcher-bearers struggled with 114 sick, mostly stretcher cases, through the 500 yards of slush to the line. Moving the sick began at 1345 hours and at 1730 hours the cholera patients were being taken to the entraining point. Fine, drizzling rain was falling and the jungle was cold and dank. At 2100 hours the train for the last of the sick had not arrived and Rampling of 'Williams' Force died. Shivering, half-naked stretcher-bearers carried his stretcher back through the 500 yards of slush for his burial in Taunzan cemetery. The train that would have taken Rampling to the hospital camp arrived half an hour later.

Having delivered the sick to the 55-kilo hospital (death house was Rowley's mental verdict), the RMO and his stretcher-bearers returned to spend the night at Taunzan before proceeding by train and on foot to the 70-kilo camp.

Mizale, previously occupied by Burmese, was terrible. The filth was so bad that Boofhead, one of the Koreans, was sickened into organising a clean-up campaign: a job quickly and thoroughly done as he did not hesitate to press the sick into service. The whole camp area was slippery mud in which the wooden clogs stuck, leaving their owners to proceed without them. Seventeen hundred and thirty men were packed into one hut about a hundred yards long, sixteen men to each thirteen by ten foot bay and occupying bunks built in tiers three and even four high.

Rowley and John Shaw decided they were reasonably comfortable with a pole-and-bag bunk each and table and stools made from bamboo. For the RAP and hospital Rowley was allocated two bays. They were already inadequate although only four days previously all the sick had been sent to the 55-kilo camp.

Rowley managed a visit, as a medical observer, to the 55-kilo hospital camp and the 40-kilo camp. Captain John Higgin, whom Rowley had seen only briefly since he had been dropped off at Mergui while the main body of 'A' Force worked on the Tavoy drome, was looking after 1,200 patients. He was worn to a frazzle with overwork and appeared quite bewildered, as if he could no longer absorb the extent of man's inhumanity. The stench of the ulcer and dysentery wards turned a stomach that Rowley thought could stand anything after the smallpox and cholera camps. The food situation was grim, owing to the Nip 'no work, no food' policy and the hopelessly ill could only be put into a small hut known as 'the death house' and from which six to eight were buried daily.

The 40-kilo camp hospital was worse, looking more like a morgue with corpses scattered among the still living in the crowded bays of the huts. There Rowley, who had kept outbreaks of tropical ulcers among 'Anderson' Force under control by his daily skin inspections and by meticulous cleaning and dressing of every small ulcer, saw tropical ulcers that extended from just below the knee to the foot, exposing all the bone and tendons. There was no treatment left for the sufferers except amputation.

About 120 of the sickest were to go back with Colonel Eadie to the 30-kilo main hospital and sixty were to be sent on to Mizale as sufficiently fit to work. Rowley went on sick parade with Colonel Eadie and discovered that the corporal deciding who should go to hospital was an old patient—

Yamamoto whom he had injected with NAB for syphilis and thereafter treated with painful hot foments. Yamamoto was a grateful patient and so Rowley was permitted to send more of the sick to hospital than the Nips had originally intended.

On returning to Mizale Rowley discovered that he had been placed on two weeks complete rest, his work to be carried out by Sid Krantz and Jim Armstrong. Despite his protests he was allowed to do no more than pay a daily visit to the hospital. It was a time of sleeping, resting, eating such food as there was to eat, and working on the plans for his house.

Apparon, the 80-kilo camp, had previously been used as cattle yards by the Burmese and moving-in was a hell of mud, manure, stretcher-bearers slipping and sliding with seventy stretcher cases, and that poor relation of a Mongolian raider, as Jim Lynch called the Storm Trooper, bashing sick, stretcher-bearers and RMO alike with a bamboo stick. He was obviously intent on showing no favouritism.

At Apparon a kitchen was not a building but a place where cooking was done. The RAP and hospital were situated on a little hill, the one dry, almost clean area. British, Dutch and a few American POWs were already at Apparon.

There was a so-called hospital camp in a valley about 800 yards from the main camp which was, in effect, a morgue with some still breathing corpses: the dump and burying ground for the sick who, according to the policy of No.5 Branch POW Camp from the Thailand end of the railway, could not work therefore should not eat and should die as quickly as possible. American, Dutch and some Australians were in this hospital camp, cared for by an American naval surgeon commander struggling valiantly with a situation beyond the control of any one man.

Although Rowley was supposed to be 'resting' he and Sid Krantz sneaked into the No. 5 Branch hospital camp to visit the American surgeon and offer aid. They found among the sick and dying a young Australian with a tropical ulcer so bad that only amputation could save his life. With the American they moved him into 'Anderson' Force RAP. With very little chloroform on hand speed was the most important factor in the operation. The American surgeon administered the chloroform. Rowley and Sid Krantz performed an amputation through the knee joint in a matter of minutes and gave the lad a blood transfusion using blood from donors with 04 blood group and the improvised giving sets from the cholera camp. Although he was in an extremely toxic state they thought he might stand a chance and after some haggling with the Storm Trooper gained permission to keep him in the 'Anderson' Force hospital.

Apparon gradually settled, literally, into mud and slush. The floors of the huts were covered with six to nine inches of mud, except on occasions when the rain was so heavy that it washed even the mud away. Apathy began to settle over the men. In three days there were as many new psychiatric cases, probably due to pellagra dementia: men who sat staring vacantly between alternating periods of intense fear, anxiety and deep depression.

Rowley had seen the will to live manifested in the smallpox and cholera camps, and now he began to learn that there could be an active will to die in men stretched beyond their limit. Apparent convalescents seemed to take stock of their situation, then decide that the ugly round of illness, convalescence, exhausting toil, mistreatment, semi-starvation and illness again was not worthwhile. They lay back on their sleeping platforms and died.

The amputee, despite neuritis, paralysis of his good leg, malaria, general toxaemia and exhaustion, fought on, fed with all the tasty titbits the RAP staff could scrounge for him. He had a ton of guts and deserved to live.

Orderly Don Booth, who had been ill since Taunzan and whose condition deteriorated rapidly during the moves to Mizale and Apparon, was another fighter.

'Looks as if he means to stay with us,' Rowley told Jim Armstrong hopefully. Jim had malaria and Rowley had taken over sick parades again and absolutely barred Jim from any RAP or hospital work.

Despite the mounting death toll the Nips demanded more and more men for work parties and worked them for longer hours. Rumour was that the Nips meant to have the railway finished by the end of September, six weeks away.

Dr Higuchi himself paid a visit to the line camps and conducted the sick parades. At Apparon he sent many of the sick out to work, despite Rowley's protests, and decreed that all remaining sick must be fit to go out within ten days. Fortunately Brigadier Varley had succeeded in having himself included in the same inspection party as Higuchi and visited Apparon on his return from the 105- and 108-kilo camps, making a thorough inspection of the hospital and observing the sick parade. A few days after Varley's visit orders came through Jap HQ that the most desperately sick were to be moved back to the line camp hospitals at the 35- and 55-kilo camps.

'The Brigadier has been remarkably successful in making representations on behalf of our sick,' Anderson told Rowley with considerable satisfaction. 'His reports will pave the way for me to plead that "Anderson" Force is a spent force and should be returned to base.'

'The Nips won't buy it,' Rowley said. 'They want the line finished and they won't send any force out of the damned jungle as long as there are men who can stagger out to work. We have to care for our sick here.'

Anderson drummed his fingers irritably. 'I know your views, Richards. We've been through this before. The Japanese have agreed to send our sick back to the hospitals. It's an important concession. We have to cooperate too. You can't keep hospitalising every man who runs a temperature or complains of a stomach ache. The fit must go out to work.'

'There are no fit,' Rowley insisted.

Anderson flared suddenly. 'You're an exceedingly stubborn young fellow. Morale in my force is low, sickeningly low, and I feel I must lay part of the blame at your door. When the men can run to you with every minor complaint, how am I to keep them on their feet? On their feet, Richards.

I've had long experience in leading men and I know that the men who stay on their feet win through. When they give up and lie down on the job they die. I'll have your cooperation or I'll make the decisions without it.'

Dismissed, Rowley went to select the first batch of sick to be sent away. He realised there was nothing to be gained by arguing with the CO. Anderson knew a lot about leading men in tough situations but little about caring for human beings in a battle against the microscopic enemy whose onslaught could only be detected in diarrhoea, colic and fevers. Rowley thought that the CO just did not believe in disease. He didn't even believe in lesions he could see for himself. Pus-filled, suppurating ulcers were of no account if a man had a leg to walk on. Only when he lost the leg would he be considered disabled.

There was no way that the soldier who demanded the triumph of human spirit over flesh and the soldier-doctor who dealt with the ills of the flesh could ever find a common ground for cooperation. It was a pity, Rowley thought, for he considered Anderson a good soldier and in battle would have followed his lead gladly.

Thirty sick—twelve 'Anderson' and eighteen 'Williams'—were sent to the 35-kilo camp by motor trucks. Rowley watched them go with mixed feelings. Once away from Apparon they could not be forced onto the line, but conditions at the hospital were so bad that he felt, miserably, that he was signing the death warrants of the men when he evacuated them.

During mid-August three more parties of sick, each numbering fifty-six, were moved to the 55-kilo camp hospital. John Higgin would do his utmost to care for them, but he was already stretched to the limits of human endurance.

On 22 August the explosive sick parade situations reached flashpoint. Out of a camp strength of 350 Rowley detailed 140 for the work parties. The Korean guard, Kobyashi, was almost beside himself with frustration and fear of repercussions from the engineers.

'More men must work.'

'No. No more men are fit,' Rowley told him in flat refusal.

Muttering, Kobyashi headed away from the RAP for a conference with Sergeant Shimojo. In the sleeping huts and around the parade ground the No Duties men sat in sullen silence with drooping shoulders and bent heads. They were treated like dogs and for the first time looked like dogs that had been beaten to the point of turning on their tormentors even though they knew that snarling and snapping at the overlords could mean their deaths.

'What do the officers care about us?' Dodger grumbled. 'They should get out of their cushy jobs and try working on the line. I'm bloody sick of them, even that doc playing God Almighty every day, saying who's well enough to work and who isn't. "You can do a couple of days, my man. It won't hurt you." Bah! We should tell 'em all to get stuffed.'

Jim Lynch, again in the grip of malaria, lifted his shivering body on one elbow. 'Shut up, Dodger. Just bloody well keep your mouth shut. The Nips starve us if we don't work. They'd just as soon shoot us.'

Exhausted, Jim collapsed back onto his sleeping platform and clenched his chattering teeth. Dimly he heard the mutterings, some in support of Dodger, some counselling caution.

'Let them not act like damned fools,' he prayed.

The CO stormed into the RAP with Sergeant Shimojo and Kobyashi at his heels.

'Richards, I understand you absolutely refuse to detail more men for the work parties.'

'Yes.'

'Only 140 out of 350 able to work! After we sent all hospital cases away only three days ago? That is ridiculous. You will send out all Light Duty personnel. Also there are officers available, six I believe. They can make up the numbers.'

'Very well, sir.'

'Now, I am asking Sergeant Shimojo to hold a sick parade, in my presence. I shall decide whether the men he chooses are fit enough.'

Rowley felt the blood tingling hotly in his face. He also felt Jim Armstrong close to his back and Old Silver's huge hand laid warningly against his elbow. Jim was right. He could not disobey the direct order of a superior officer.

By mid-afternoon two of the officers and two men who had been chosen by Shimojo were sent back from the line too sick with malaria to work. Among the sick Rowley had kept in his hospital there were four deaths that day, including Chiddy who had been continually ill since Tavoy but, like Chauncey, was a game battler.

For the first time in the history of 'Anderson' Force there was difficulty in obtaining volunteer pall bearers. Group loyalty had taken a beating.

Unexpectedly and at very short notice Rowley and Sid Krantz were removed from the jungle hell and, for a few weeks, from the unequal battle against starvation, disease and brutal disregard of human suffering. The order came from Colonel Nagatomo but the motivating force in the background was Brigadier Varley.

POW Standing Order 17 issued from Colonel Nagatomo's headquarters and dated 2503/8/23 stated:

> In order to alleviate temporarily the task of the doctors who have continually worked with the utmost zeal and give them the opportunity to foster their health, there will now be a temporary change of POW doctors as mentioned below:
>
> Major Hobbs to change with Major Krantz.
> Major Chalmers to change with Captain Richards.
> Captain Higgin to change with Captain Anderson.
> Captain Huls to change with Captain In't Veld.

Rowley received his orders with a mixture of feelings that verged towards dismay. He feared for the welfare of the men whose survival had become his immediate reason for existence, and could not imagine how doctors from base hospital, skilled and dedicated though they were, would cope with the conditions in the deep jungle. As for himself he had never considered any other outcome than that he would survive the line and the war and eventually set up his practice and build the house which, thanks to John Shaw and Melbourne architect, Captain Stuart Handasyde, had now taken shape in a set of beautifully drawn plans. He felt he was quite fit to go on, but as is usually the case, he was not his own best physician. Jim Armstrong was quietly relieved and Roy Whitecross, working in the RAP as his leg ulcer had broken out, voiced the general opinion, 'If the doc had kept on going much longer we would have had to bury him too.'

Retpu was like heaven's antechamber, or at least an approximation of the almost dreamlike memories of a life before the fall of Singapore. It was a well-run camp with a generously stocked canteen, regular concerts, and an atmosphere of relaxation. Major Ted Fisher was in charge, spruce in well-laundered khaki trousers and shirt. Just as Rowley arrived he saw Chauncey, on a truck and ready to rejoin 'Anderson' Force. Rowley had sent Chauncey back to Retpu, sure that the haven of a hospital was his only chance of survival.

'Where are you off to, Chauncey?'

'Back on the job, sir. They say I'm too fit to stay here now. I feel fine.'

Rowley's medical duties were no more than looking after 200 convalescent dysentery cases, rounds which he could do in thirty minutes. Retpu dysentery ward was child's play for the RMO of 'Anderson' Force. For the rest of the day he could eat, sleep, swim, play chess or cards, read and attend evening concerts. A pleasant euphoria invaded his overworked mind and body, yet he could not see the healthy men around him without thinking bitterly of the men on the line. So many could have been saved if the fit had been sent out to relieve them. He could not assess where the responsibility lay for failure to rotate the healthy with the desperately sick.

Before his thirty days were up Rowley was feeling fit and impatient to be back with his force and was agitating to be returned. He was also scheming to liberate some supplies of cholera vaccine to take with him. There was no knowing what sort of camps his men were inhabiting while he was away. He rejoined 'Anderson' Force at the 131-kilo camp, called Little Nike by the POWs, on 16 October, to find that things could and had become worse.

For weeks his force had been in a filthy pest hole at the 116-kilo camp where sleeping huts were built in a gully and where the primitive sanitary habits of the natives were lamentable as usual. The watercourse had washed down mud and faeces into the 'Anderson' Force huts. Diarrhoea and dysentery had increased and the malaria cases, weakened by diarrhoea and chilled by working in the rain, were barely over one bout of fever before suffering relapses. That human bodies had withstood so much punishment for so long was a triumph of the spirit.

On top of the unhealthy assaults of the jungle, the Japs had been literally working the men to death in an effort to finish the line. Guards had been continually blitzing the sick parades and the hospital. The line, Rowley was told, was virtually finished, only a little tidying up to be done.

Anderson informed Rowley with satisfaction that the health of the men had improved since they had moved to Little Nike, a clean camp. Fewer men on sick parade were ill enough to be kept in camp and larger work parties had been provided with the result that relations with the Japanese were much improved.

The sick parade figures produced by Allen Taylor seemed to support Anderson's claims.

'It's been "Anyone who can walk can work". Our protective rest system has been blown to hell,' Taylor said tersely. 'Men are half-carried to work by their mates.'

Major John Chalmers lifted his hands in a gesture of weary resignation. 'What can we do? The Nips take as many as they want. I protest but they just don't hear.'

Jim Armstrong was ill, and so was Bob Kemp. Both were apologetic, feeling unjustified guilt because they had collapsed under pressure. Chauncey was in hospital again, looking older and thinner, a wizened monkey with patient eyes. He lay quietly, uncomplaining, smiled slightly when he saw Rowley. There was a look of death about him.

Rowley conducted evening sick parade. There was a lad who complained of a sore throat and shivers. Immediately Rowley put him on Light Duties and told him to report for work in the RAP as they were short-staffed with Armstrong and Kemp ill. The lad thought it was decent of the doc to offer him a job in the hospital, especially as there wasn't really much wrong with him. Of course his feet were sore but he hadn't mentioned that this time. He had used it once before with Rowley but had been sent out on a work party just the same with a laconic, 'Sore feet? You can still walk to get your tucker I notice.' Seemed the doc didn't worry much about sore feet, even if a bloke did have to tramp for kilometres in the mud beside the rotten line without boots for his feet. Seemed a sore throat was the thing to toss the doc.

Roy Whitecross had no idea that the doc had not been tossed at all. The sore feet had been noted, as well as the sore throat and shivers, and before that the trouble with vision, and almost a year before, at Thanbyuzayat, the muscular weakness and the *stomatitis* at the corners of the mouth. The young soldier was sick without knowing it, a recurrence of pellagra, probably early second-stage, and Rowley was prescribing the only treatment available to him to keep the boy alive and sane.

The lad was given the task of dressing tropical ulcers. Chalmers watched the care with which Rowley's unofficial and official orderlies patiently picked the tiny fragments of rotten flesh out of the suppurating holes. Apparently Rowley didn't use the spooning-out treatment favoured by the Dutch doctors. Ulcers! Chalmers had treated and smelt so many of them at the HQ hospital and had amputated limbs where inches of diseased bone were exposed after the rotting flesh dropped away. Tropical ulcer cases for amputation came in from all along the line, but none from 'Anderson' Force. Chalmers had thought about that a great deal during the last few weeks. It seemed that Rowley Richards' methods worked.

When Rowley arrived at Little Nike there was one patient with cerebral malaria in the hospital. His history of malaria went back to Tavoy and over the last six weeks he had suffered recurrent attacks every fortnight. Inevitably the disease had attacked the brain.

In such cases the delirium and hallucinations rapidly became worse and the patient, deriving unnatural strength from his terrors, would be continually active, attempting to get out of bed, flee the hospital, even the camp if he could elude the orderlies and the guards. But he could not escape the jungle. For a while he might thresh around in undergrowth that was a woven mat of bamboo and vines between the vast trunks of the teak trees. He might collapse and die before the search parties found him. If they got him in time he would be brought back to the RAP exhausted and semi-conscious.

He would then lie in that state for anything from two to four days, after which he would either recover or lapse into unconsciousness . . . rapid, gasping respirations would develop and he would slowly and quietly die. In the unlikely event that he lived there would be an inquiry into the 'break', but the Japs would not have him executed as an escapee. They would find him mad. They had a superstitious dread of madness, and a mad ghost coming back to haunt them was more to be feared than a sane one.

Shortly after the train carrying the Japanese VIPs to the line-joining celebration had passed through the camp, the man with cerebral malaria lapsed into semi-consciousness. His breathing was fairly steady. Perhaps he would live. He lay quietly hour after hour, not stirring when news was brought of Corporal Rooney's death, nor later when Bombardier Stanley was brought into the hospital.

Just before dawn forty-eight hours later Rowley was wakened from a light sleep by the sound of short, rapid breathing. Pinky was already awake and with the patient, who did not regain consciousness. The gasping breaths began to rattle and he died at first light.

Following the joining of the line near Three Pagodas Pass, all work on the railway ceased and the men were allowed to rest and clean up the camp. Rowley was given permission to go to Little Nike for supplies. He and Jim Armstrong set out on foot for the village, escorted by a Nip guard. The Nip, now that the engineers had left, was relaxed and almost friendly. At the river between the camp and the village he stripped off to his G-string, made a bundle of his clothes, which he carried on top of his head, and forded the river through water that was knee to hip deep. Reaching the further bank he began to bathe and cheerfully indicated that Rowley and Jim could do likewise. Jim, weak from his last bout of malaria, was content to lie under trees and Rowley stayed with him examining the Thai money that Sid Krantz had brought back from base. It was the first Thai money they had received and could be spent in Little Nike, where the villagers would not accept Burmese money. When the guard had finished his bath, dried himself on his G-string and dressed, they tramped into the village. It consisted of a series of houses dotted at infrequent intervals along a road about a mile and half long, but to the two white men who had been prisoners for so long in the deep jungle it was a thriving town.

'It's bloody civilisation,' Jim gasped. 'Watch out crossing Main Street, doc. Bushies like us could get knocked down in this traffic.'

Lumbering elephants were being ridden along the road by their mahouts, with noisy shouting; there were rickshaws and people on foot wearing Western clothing, the men in grey slacks and sports jackets with—wonder of wonders—collars and ties, and women in skirts and blouses with shoes on their feet and even carrying handbags. The Thai police, strolling through the market area, wore uniforms with gold braid and brass buttons that would have been a credit to any theatre commissionaire.

Rowley investigated the shops which were set up in the front rooms of the dwelling houses and discovered that some of them were even furnished with glass display cases. Milk and bananas were plentiful and when the friendly Thais discovered that his purchases were for the sick they produced some sago flour and *gula malacca.*

The colour patch on Rowley's faded hat caused some excitement and a group of villagers clustered around him pointing to the patch, managing to convey to him that they had seen other Australian officers wearing similar ones. One smiling girl, magnolia-skinned and dainty, even remembered some of the officers' names, but Rowley did not recognise any of them.

'They must have been members of the 2/10 Field Regiment,' he told Jim. 'I guess they were sent out of Changi after us and worked on the Thai side of the line. If they send us back to Singapore, there'll be some stories to be told.'

Back in the camp there were many rumours and surmises as to what would happen to 'Anderson' Force next. Some said they would move to the 116-kilo camp, to Bangkok, to Saigon, or to Singapore. But no one knew anything definite. Colonels Nagatomo and Nakamura inspected the camp and following their visit the guards were all blind drunk for twenty-four hours and there was not one guard on duty.

'They've no right to neglect us like this,' Scippy Maher complained to Jim Lynch. 'I don't feel right without the bloody Nips buzzing around abusing and bashing us.'

But the celebrating and relaxed attitude of the guards did not last. Within a few days they began to get jittery and would not allow the men to gather in groups of more than three or four. There seemed no reason for the tightening of control, unless the Nips thought that the prisoners might start plotting a mass breakout now that they were no longer working on the line. A silly idea since the men had neither the strength nor the will to tackle the jungle. The No Duty sick figures, at an all-time low during the last push to finish the line, were mounting rapidly.

This was a phenomenon that at first surprised and puzzled Rowley. The general health should be improving now that the pressure was off, yet the malaria and pellagra cases were more severe. It did not seem to add up as part of the survival equation. He finally reasoned that men who drew on almost superhuman reserves of energy when under stress were now completely exhausted, like marathon runners who collapsed once they breasted the tape. They were going to need to reorganise psychologically. Some new challenge might be necessary.

Old Chauncey, helped by the extra food, was battling on and looked as if he would beat the reaper, but John Shaw was down with a severe attack of dengue fever, suffering from crippling headaches and running a constant temperature, and courageous, uncomplaining Jim Lynch was again in the grip of malaria. He tried to make light of his condition and minimise the number of attacks he had suffered while Rowley had been at Retpu, but his weight was down to nine stone five and he looked haggard.

The cheerfulness Lynch had maintained throughout the horror months on the line came now only in rare, weak flashes. There was an issue of clothing and he joked to Rowley, 'I now have one green hat, too small, and one green coat, too large. I'm the picture of what the well-dressed lawyer should wear on safari.'

'You're the picture of one lawyer very sick with malaria,' Rowley told him. 'Stop cracking hardy, Jim. You're on No Duties until further notice. Just eat and sleep.'

'If it helps my temper Scippy might find me easier to live with,' Jim Lynch admitted. 'The last week my mental state would have made a fascinating study for a psychiatrist.'

Rowley gave him a hard look. 'It might be just as well to hospitalise you for a few days.'

'No, please. I'd imagine I was sick if you did that. I'll be good and rest as the doctor orders.'

Ari, the Storm Trooper, had other ideas about Lynch, who always looked him in the eyes and grinned at his abuse. Ignoring the No Duties decree of the RMO he ordered Lynch onto night guard duty for his hut: the POWs also had to mount guard at night. His relief was to be Scippy Maher.

At midnight when Jim's turn of duty was ended Scippy was still sleeping. Quietly Jim left his beat outside the hut and went to wake his mate. Scippy had just rolled off the sleeping platform and was heading for the door when the two heard Ari shouting outside the hut. The commotion he was making awakened all the men and there were sullen, angry mutterings as Ari herded Jim Lynch and Scippy outside.

Ranting steadily, Ari dispatched a nervous young Korean guard to awaken Colonel Anderson and Bill Drower and demand their presence.

Bill Drower told Rowley later that Ari seemed to have lost all control. Refusing to listen to the explanation that Lynch had merely left his post to wake up his relief, he began slapping both offenders, the Colonel and Drower. As the violence of the beating increased even the Jeep became nervous and tried to restrain his fellow guard. Ari's mind seemed to snap and he rained a flurry of blows on all four defenceless men until they collapsed. Anderson hit his head on a post as he fell and was knocked unconscious. Ari would have continued putting in the boot as they lay on the guardhouse floor, but the Jeep persuaded him to leave and sent for Rowley, who was in the hospital administering quinine intravenously to Gunner Joe Field, 2/15 Regiment, a critical malaria case.

The CO and Bill Drower recovered quickly and needed no treatment, and Scippy was fit enough to go back to his hut after his contusions had been bathed and treated, but Jim was slow to recover, dazed and too weak to stay on his feet. Rowley kept him in the hospital section for the night but he appeared better in the morning and asked to be allowed to go back to his hut.

After the bashing the temper of the men was sullen, with an undercurrent of anger coming close to violence for the first time since the incident, months before, when Bill Drower had been jumped by guards and kicked over the matter of the cigarette case. The Storm Trooper was removed from the camp by the Japanese and the Jeep took over command of the guards.

Chauncey, the grand old battler, suddenly deteriorated and died.

Death also won the battle for Joe Field but John Shaw, to Rowley's relief, was recovering from his attack of dengue fever. John had turned forty-one at Apparon and was an old man in terms of average ages on the line. Rowley worried that the fine mind of his friend might become the victim of cerebral malaria.

It was, however, the fine mind of Sergeant Jim Lynch that was the next victim of cerebral malaria. He had been very quiet but had seemed to be recovering from the bashing and had asked to be put on Light Duties as he liked to keep occupied. A week later he suffered another attack of malaria and was supported into the RAP, shivering, between Scippy Maher and Rusty Irwin. Jim Armstrong sat beside him through a day and night of sweating and rigors, then the attack seemed to subside but the fever continued.

Rowley administered intravenous quinine and the sulphonamide, M & B 693, but there was no improvement. From delirium Jim Lynch lapsed into coma. His face was grey, the skin without tone, staying up in little puckers when Rowley tested it between finger and thumb. His rapid, gasping breathing became steadily weaker and he quietly died.

'He had all the courage, will and stamina a man needs to survive and yet he lost out when the worst was over,' Rowley said sadly to John Shaw. 'When it comes to the crunch neither medicine nor willpower can help. It's all up to Providence. All we can do is keep battling on and believing that it's not going to be our turn yet.'

Jim Lynch's was the fifth death from cerebral malaria. There was one death from pellagral dementia, but the ulcer cases were improving and Rowley noted with satisfaction that there had still not been one amputation among the men of 'Anderson' Force.

Statistics were received from headquarters detailing the number of deaths of No. 3 Branch POW Camp up to 31 October 1943, broken up into various forces, their strengths and number of deaths. With a force strength of 731, 'Anderson' Force had had sixty-nine deaths, representing 9.3 per cent. 'Williams'

Force had 128 deaths out of a force of 824, representing 14.5 per cent. 'A' Force as a whole had 289 deaths, or 9.6 per cent of a total force of 3,000.

On 20 and 21 October the Japanese organised two special events; the first a memorial service for those prisoners of war who had died during the building of the Burma-Thailand railway, which took the form of the dedication of the cemetery area and the erection of an eleborate cross carved at POW Headquarters and sent out to the point where the lines had joined. The second was a celebration of the first anniversary of the founding of the Burma-Thai prisoner-of-war camps. At the memorial service a speech of condolence to the souls of the dead, prepared by Lieutenant-Colonel Yashitada Nagatomo, was read.

> As the first stage of the railway construction has now been completed I have, on this day of commemoration, the honour of taking the opportunity of consoling the souls of the Prisoners of War of the 3rd Branch, numbering 655, who have died in this district during the last year.
>
> In my opinion it is a virtue, since ancient times, to pay homage to the souls of those who have died in war, even though they may be enemies. Moreover you were under my command and have endeavoured to work diligently in obedience to my orders, while always longing for the final repatriation to your own countries, once the war was over and when peace was restored.
>
> I have always done my utmost to discharge my duties conscientiously taking responsibility for you all as your commander. Now you have passed on to the other world, owing to inevitable prevailing disease and epidemics and to the indiscriminate enemy bombings, I cannot see you in this world any more. Visualising your situation and especially that of your relatives and families, I cannot help shedding tears, sympathising with your unfortunate circumstances.
>
> This tragedy is a result of the war, however it is owing to fate that you are in this condition and I consider that God has called you here. However, today I will try to console your souls and remember for you, in my capacity as your commander, together with other members of my staff, by dedicating a Cross and placing a wreath in your cemetery.
>
> In the very near future your comrades will be leaving this district, consequently it may be impossible to offer prayers or place a wreath in your cemetery for some time to come; but undoubtedly some of your comrades will come here after the war to pay sincere homage to your memory.
>
> Please accept my deepest sympathy and sincere regards and may you sleep peacefully and tranquilly.

The speech read on the following day to commemorate the establishment of No. 3 Branch on 15 August 1942 and the arrival of 'A' Force on 2 October 1942, extolled the great feat of building the railway in florid prose:

> Under difficult conditions, such as lack of transport facilities, imperfect and insufficient accommodation, privations, rations and medical supplies, we have

exploited untrodden jungles. Under the burning heat of the tropical sun and dull torrential downpours of rain, we have achieved this epochal and brilliant feat in this period of time with the inflexible and indefatigable energies of those who wielded the pick and shovel.

This achievement reflects great credit upon us and must be attributed to the fact that each of you has been zealous in doing your own respective work, grasping my mind and myself observing my instructions of various times and my own rules since the establishment of No. 3 Branch. I extend to you many thanks for your labour, with the deepest regards.

Now the organisation of this Branch is confronted with new dispositions, which have developed in accordance with the change of the construction work, all Prisoners of War then ought to readjust their minds to the aims of the commander and to be absolutely obedient. The members of the Japanese staffs should realise the POWs' actual conditions and should not miss any opportunity of contributing to their betterment. Both sides should try to understand each other. In this way we are working together for the prosperity of this Branch. On this occasion I strongly emphasise the abovementioned items.

Happily let us celebrate this memorable day by having a very pleasant and cheerful time to everyone's heart's content.

The practical celebration of the event was a grant of extra food and a special canteen service and the arrival by truck from Moulmein of two pigs each for 'Anderson' and 'Williams' Forces. They had been purchased by Jack Varley, the Brigadier's son.

There was also a concert given by a combined concert party selected from the various camps of No. 3 and No. 5 Branches.

Nagatomo's statement that the organisation of this branch was confronted with new dispositions started a new flurry of rumours and speculation over what was the next job in store for the prisoners. The rumours were finally resolved into definite information: the thousand sickest were to go to a new hospital camp at Tamarkan near Bangkok; 6,500 were to go to a rest camp; maintenance parties of 3,000 were to be left along the railway line.

The POW camp at Tamarkan in Thailand was a paradise to men who had known the filth, slush and disease of jungle line camps. It had been occupied by British prisoners under the command of Colonel Toosey, an exceptional officer, and was known by repute for its cleanliness, its kitchens, canteens, gardens, duck yards, pig pens, cattle yards, playing fields and concert parties. With military police and house guards, it was run on the lines of peace-time barracks.

The camps at Tamarkan and adjacent Kanchanaburi had provided the work force to construct a ten-span steel bridge over the wide and fast-flowing Kwa Noi River. The bridge was protected by anti-aircraft guns on its approaches and the peace of the camps was regularly shattered by practice gun drill.

Rowley was one of an advance party of fifty which, under Major Don Kerr and Major Rupert Barraclough, took over the camp for Nos 3 and 5 Branch

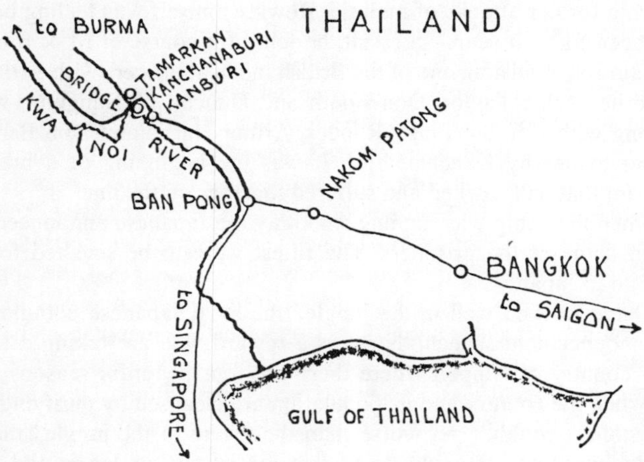

prisoners. A rear party of Colonel Toosey's camp, mainly British with a few Australians, was there to welcome them. So were Nip guards who searched their persons and kits thoroughly and found and confiscated the first part of Rowley's laboriously kept diary which was carefully secreted in the bottom of his kit. The second part, beginning with Christmas at Hlepauk and which was kept quite openly with current Jap medical records, was not found. Rowley was relieved that the offending record was merely confiscated. Severe punishments, even execution, could be meted out by the unpredictable Japanese for keeping accounts that detailed war crimes.

Major Arthur Moon, a Sydney gynaecologist of the 2/2 Casualty Clearing Station from the Middle East and a POW in Java, was in charge of the hospital and made the entrance of the advance party into the camp a memorable one with a feed of meat, vegetables, eggs, even savouries, and cakes and some money to make purchases from the canteen. For days the jungle survivors drifted through the fleshy delights of Tamarkan with a feeling of unreality.

The first of the main body from the 55-kilo camp arrived at Christmas time and the last from Nike were at Tamarkan by 13 January 1944. There were then 3,000 Australians and 300 British from Chungkai in the camp under the command of Colonel George Ramsay of the 2/30 Battalion.

Rowley was officially Liaison Medical Officer and Don Kerr was nominally camp 2 I/C. Within a month, in conjunction with Major Colin Cameron of Western Australia, the Hygiene Medical Officer, they had a smoothly running camp and the health of the troops was improving visibly, although there were some who had been so badly affected by jungle conditions that they continued to deteriorate. They were the first, Rowley wondered, of how many who would return to civilian life old, sick men and die before their time because of the ill-treatment they had received.

Jim Armstrong was not as well as he should have been for his age and was having further attacks of malaria. Rowley himself was feeling better than he had been for two years—a result, he felt, of a course of M & B 693 given by Captain Jim Roulston, one of the British medical officers with Arthur Moon. His orderlies Allen Taylor, Don Booth and Duncan McNaughton were with him, along with Jim, but Pinky Rhodes, Arthur Harris and Don Bertram had been sent to nearby Kanchanaburi. It was the beginning of a breaking up of the team that had worked and suffered together on the line.

Just when the camp was running smoothly the Japanese announced the next move for some of the prisoners. The fittest were to be selected for a work party to go to Japan.

'You have worked well in the jungle,' the local Japanese commander said in his announcement speech. 'Now, as a reward, you are taking a trip to the glorious country of Nippon where there are four beautiful seasons including winter, when the country and its population are cleansed by purifying snows.'

'Well, things couldn't get worse than they were in the jungle,' the Dodger opined to his mates. 'I could do with a tourist trip to Japan and a chance to pick up some curios. There might be a bit of money to be made.' He was delighted when he was chosen among the fit.

The Japanese made the selection, giving the Australian officers very little say. Jim Armstrong and John Shaw were passed over as too old. Rowley was chosen . . . mainly, he thought, because of his youth and his known intractableness as a medical officer on the line.

Realistically facing the fact that there was the chance of Japanese transports being sunk by American submarine packs in the South China Sea, Rowley gave all his medical records and main diaries to Jim and John Shaw to take home if they could, but he intended to carry with him a six-page summary of his diary and a record of sickness and food in graph form, together with his pay book and other personal possessions, in an improvised waterproof wallet. John Shaw used the false bottom of a billycan as a hiding place for the diary and Jim concealed the main reports and papers in his medical kit.

The Japan Party was divided into six kumis, numbered 35 through 40, each of 150 men plus one officer of the rank of captain or lieutenant and one member of the Australian Army Medical Corps. The kumis were then made into groups of three with the AAMC officer attached to the first kumi of the group, an AAMC NCO to the second, and an AAMC private to the third. Captain Arthur Sumner, 2/29 Battalion, became officer commanding Kumi 35, with Rowley as medical officer. Lieutenant Les Stupart, 2/10 Field Regiment, commanded Kumi 36 with Corporal Len Coon as medical orderly.

On 27 March 1944 both these Kumis left Tamarkan for Kanchanaburi on the first leg of a journey to Saigon for embarkation to Japan. It was to take three months and much shuttling around in closed cattle trucks and a small river steamer, with a lengthy work period in Saigon, before they were to embark.

The first attempt at embarkation was made at Saigon, but after twenty-four hours on cargo vessels of about 6,000 tons they were disembarked and sent back to Saigon. American submarine activity made the waters too dangerous for Japanese shipping. By this time they had been joined by Kumi 38, with John Chalmers as Medical Officer, and by Brigadier Varley who was to travel with the Japan Party as a supernumerary on his way to the Senior Officers' POW Camp in Formosa. Somehow the Brigadier had been overlooked when senior officers had been sent to Formosa after capitulation as part of Japanese strategy for demoralising POWs by separating senior officers and men. Varley had been the only senior officer left in the jungle—a fortunate oversight for the men of 'A' Force, whom he had done everything in his power to protect.

The shuttling around finally took Kumi 35 to Singapore where they were put to work on the building of a graving dock on the south-west tip of the island. They were quartered with Kumis 36, 37 and 40 on Paulau Damarlaut, a small island about a mile from the dock and which Rowley was prompt to nickname Jeep Island as the Jeep was the Nip guard in charge of their camp.

The Jeep revelled in his appointment, reigning supreme over his handful of Korean guards and a working party of 600 Australian POWs and out of the reach of Lieutenant Naito, Sergeant Shimojo and other Japanese soldiers who considered Korean soldiers as mainly criminals who enlisted to avoid serving gaol sentences and despised them almost as much as they did prisoners of war.

The Jeep's 2 I/C was Greenpants, a third-class private who wore oversized pea-green breeches, who also revelled in his exalted status but had had no experience in handling the prisoner camp staff like Captain Sumner and Rowley. As a result he misjudged Rowley's calibre, and Rowley badly miscalculated in his handling of Greenpants with painful results.

It happened during a sick parade.

'Do you want a hand with this chap?' Arthur Sumner asked Rowley.

'No thanks, I'll be all right.' Rowley, with many jungle sick parades behind him, was confident. 'I've seen him in action on working parties but this is his first sick parade. So I should be able to handle him.'

The parade got under way with the usual bargaining over which men should work. The first was a malaria case and Rowley told Len Coon to mark him No Duty two days. Greenpants, who had seen Sergeant Shimojo in action during Burma Blitz sick parades, immediately argued: '*Byoki nai—shigoto—ima*' (Not sick, work now).

Rowley palpated the man's upper left abdomen, with a wink to the patient to pretend pain, then suggested that Greenpants prod the enlarged spleen himself. The spleen was not, in fact, enlarged, but the man writhed convincingly and Greenpants, afraid to lose face by admitting that he could not feel the spleen, agreed to Light Duties in camp—*Shigoto campu.*

So the parade continued until they came to Gunner Maurice Barkley, 2/15 Field Regiment, who had been drafted into Kumi 41 which joined Kumi 35 at Paulau Damarlaut. Maurice was recovering from a severe bout of malaria with associated dysentery, and Rowley, knowing he had almost succumbed to severe pellagra on the railway, was determined to protect him. When Greenpants became stubborn Rowley tried his ace tactic.

'This man *takusan byoki*—he *shigoto*—he *shinda* [the man very sick—he work—he die]. *Sensu awari—ichi ban Australia* speak to me and ask why this man *shinda* [after the war the Australian No. 1 will ask me why this man died].'

Faced with similar tactics Shimojo or the Storm Trooper would have withdrawn their demands. But the inexperienced Greenpants, caught between fear of reprisals and fear of the engineers on the graving dock who were demanding more workers, lost his head. He lashed out at Rowley with boots and fists. Attracted by his screaming of curses, guards raced down with fixed bayonets to prevent any action by the prisoners while Greenpants gave Rowley the worst beating of his experience. Occasional cuffs and clouts on the line had been nothing compared to this.

As Rowley became more and more dizzy with each blow he heard voices behind him advising—'Stay on your feet'—'Don't let him get you on the ground'—'The bastard will put the boot in.'

Finally Greenpants simmered down and the sick parade was dismissed. All Rowley wanted to do was get away on his own and lick his wounds. Brushing aside offers of assistance from Sumner and the men, he wandered down to the nearby beach and knelt to bathe his bleeding lips. The high salinity of the warmish tropical water stung the cuts and as he clenched his jaws against the pain he became aware of small fragments of tooth in his mouth. The two upper front teeth and several others were broken along the edges. He had certainly been too confident and overplayed his hand.

His fingertip examination of the teeth was interrupted by the urgent tone of a little guard telling him that the Camp Commander wanted him at the guard house immediately.

'That rotten little bastard wants his cut too,' Rowley mumbled. He braced himself as he entered the guard house, remembering that even bathing in the sea without permission was forbidden.

The Jeep rose from his place at the end of a table, bowed and pointed to a seat. The table was set with a bowl of fruit, a pot of tea and two cups. Rowley was reminded of the occasion in the jungle when the drunken Naito had given a party. But the Jeep was not drunk. He was performing an act of courtesy to a courageous enemy. As he offered tea with a bland smile he said, 'You like Colonel Williams. He very foolish and must be punished, but he very brave man. All Japanese soldiers respect brave man.'

Rowley left the guard house with the fruit from the bowl as a gift he had not been allowed to refuse and the feeling that the Jeep knew and approved that he would give it to the sick. The Oriental mind, he thought, had more layers than an onion.

But the eventful day was not over and before nightfall Rowley was to discover that he, too, was capable of some complicated and devious thinking. To relieve his feelings he sought refuge in the uncertain privacy of his billet and began recording the bashing incident in his diary. Just as he was returning the diary to its hiding place in his haversack he was disturbed by the voice of the orderly sergeant calling that Corporal Coon wanted him in the RAP right away. The urgency of the tone sent him on his way at a run.

Corporal Gorlick and a mate had been brought in, Gorlick unconscious and the mate vomiting violently. It appeared they had found a dead crab on the beach and had eaten the flesh. In half an hour, because it was impossible to wash out his stomach while unconscious, Gorlick was dead. His mate, after having a pint of sea water pumped into him as an emetic while still conscious, got rid of the poisoned food and began to recover.

Rowley was sure that botulism in the flesh of the crab had killed Gorlick, but when he reported the death to the Jeep he described it as a case of cardiac beriberi and stressed, untruthfully, that Gorlick, had been one of the men that Greenpants had wanted to send to work. The Jeep was so impressed that cardiac beriberi was an undetected and rapid killer that he insisted on Rowley checking his heart. The examination served the purpose of convincing the Jeep that a doctor's stethoscope was a very important and necessary instrument.

Corporal Gorlick, Army Provost Corps, was buried with more ceremony than had been possible for many who died on the railway, with all the kumi in attendance, except for men watching the kitchen fires. The grave was on a little rise which overlooked the sea towards the south, the direction of Australia. As the body was lowered, Rowley suddenly realised that the opportunity had come for him to bury his summarised diary and other records which he felt should be preserved in the event of his being lost at sea en route for Japan.

Before lights out, by the flickering of a candle improvised from wax stubs discarded by the Japanese, melted into a bamboo tube and supplied with a wick of teased string, he removed the summary of his diary from his wallet and made a summary on one sheet of the thinnest paper he could find. He rolled up the summary and inserted it into the tubing of his stethoscope, now sacrosanct as far as the Jeep was concerned. The other records he inserted into the bottom of a beer bottle salvaged from the guards' garbage. He loosely packed the remainder of the bottle with paper and sealed it with candlewax.

The following day, during the midday meal break, he slipped away from the camp and returned to the grave at the head of which a hole had been dug for the erection of a wooden cross. He buried the bottle in the hole, upside down so that if water seeped in it would not reach the records. Then he tamped three inches of dirt over it.

The Jeep himself supervised the erection of the cross and when he indicated that it was too high and wanted the hole deepened Rowley held his breath. But fortunately the Jeep agreed to Arthur Sumner's suggestion that several inches be chopped off the upright with an axe that was at hand rather than have someone return to the camp for a spade. Finally Rowley made a map of the location of the grave, using an outcrop of rocks at the eastern end of the beach as a reference point. That he also inserted in the stethoscope.

In less than forty-eight hours he had won the Jeep's regard and established cardiac beriberi as a suitable complaint for men he wished to keep on the sick list, turned the tube of his stethoscope into a safe hiding place, and hidden his diary where it would be found by the War Graves Commission if he did not survive.

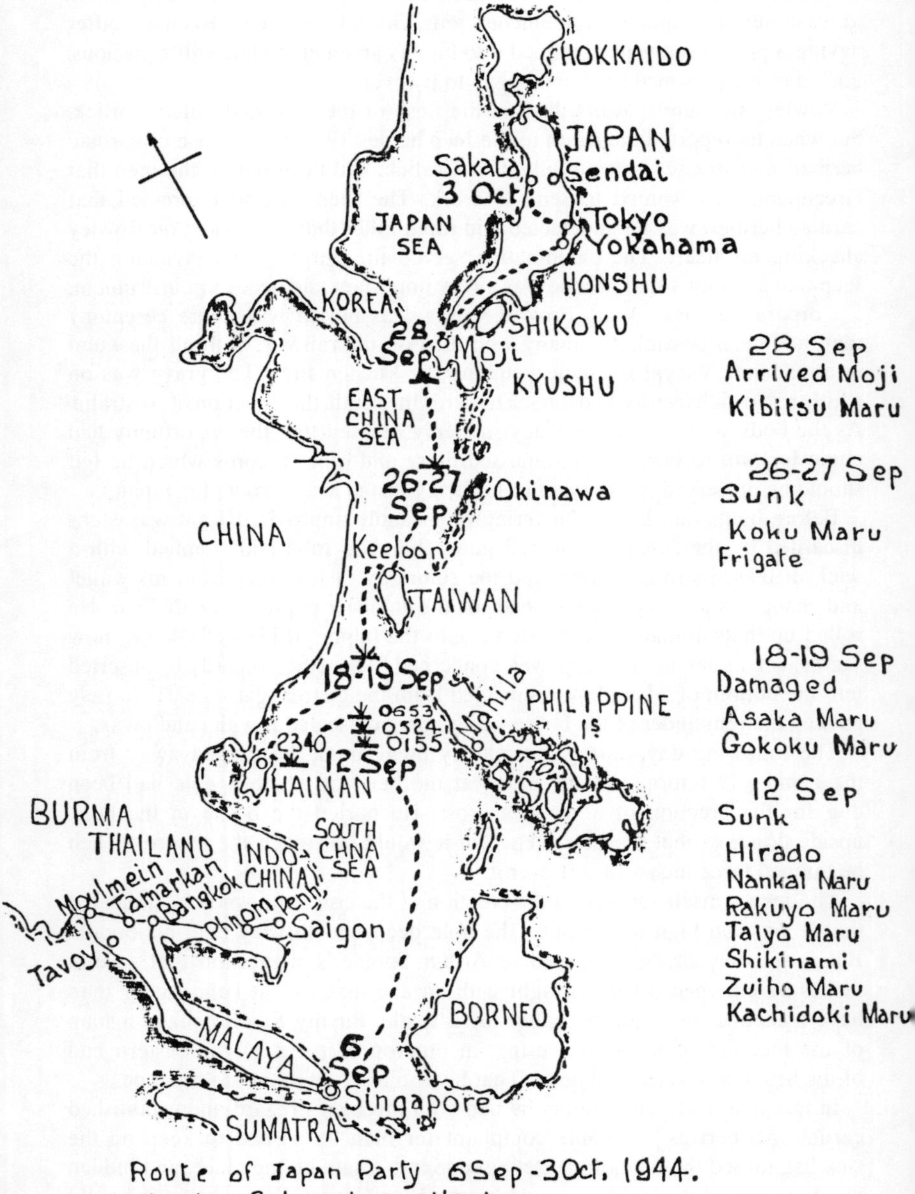

Route of Japan Party 6 Sep - 30 Oct. 1944.
Submarine attacks.

10

The South China Sea

It was stinking hot in the hold of the *Rakuyo Maru*, and stinking with the stench of unwashed, sweating human bodies. In the semi-darkness, the hatches being battened down as the ship was at sea, Rowley could make out the nearer portion of the platform, some twenty feet wide, that was built around the hold at a height of four feet from the deck so that a double layer of bodies could be accommodated. Men were packed in under the platform and on it, without sufficient room for anyone to lie stretched out. Some sat with their knees under their chins, others lay on their sides or on their backs, with their legs drawn up. More men were curled or hunched on the open deck space of the hold immediately beneath the hatches. The Australian prisoners occupied the port side and in the dimness to starboard were the British with their own officers and system of organisation—if anything could be called organisation under such conditions.

Rowley picked his way among the bodies, checking on his sick and alerting men to get up on deck for sick parade since day and night merged in the hold, time was losing meaning, and they could not know when to go topside unless they were told. Nearby he could see Captain Arthur Sumner, bent over, counting off men to go up for a breath of air and their turn at the latrine.

At least there was some room to move around. When they were embarked at the Singapore docks on 6 September 1944 the guards, acting on orders, had tried to cram all the prisoners into the hold. He remembered being one of the crowd being pushed down into the hold with the assistance of the guards' hands, boots and rifle butts. There had been standing room only once the platform had been packed and men who fainted in the furnace-like heat had to be handed out over the heads of the others back on to the deck. Major John Chalmers, Rowley, Captain Arthur Sumner and Lieutenant Les Stupart had wormed their way out of the crush against the crowd that the guards were still trying to shove in, and had reached Lieutenant Yamada who was in charge of the Japan Party. Arthur had shouted his protests over the din, supported by John and Rowley as the medical officers. Yamada, for once, had been entirely in agreement with them. He could see that it was not possible to confine over 1,300 men in the hold, as the captain wished, even for the period the ship was in Singapore, let alone for the entire voyage to Japan.

Rowley had caught a glimpse of Brigadier Varley on the crowded deck and had then lost sight of him, but he had the impression that Varley had reached the captain with a protest as that exalted person—actually a rather scruffy looking fellow in high-buttoned blue jacket and blue trousers that looked in need of a brush and press—appeared while Rowley's group were pressing their complaints to Yamada.

Yamada had insisted that the captain see conditions for himself, but before an inspection could be made some of the men already in the hold had to be brought up again so that the captain would have room to move. Finally the captain agreed that 200 men would be accommodated on deck. Yamada had acquiesced when Rowley and John pressed that the sick be given first preference in the use of deck space.

Since the convoy had left the Singapore Roads many more than 200 men slept on deck at night. After dark Rowley, Sumner and Stupart slipped below and sneaked more men up the narrow steps that were the only entrance and exit once the hatches were down. They slept not only on deck but above deck, roosting in every nook and cranny for'ard. The aft deck was off limits to POWs and reserved for the passengers including, Rowley thought, from the glimpses he had managed, a number of troops and some of their prostitutes.

He had not seen Brigadier Varley since embarkation and surmised that he would be under guard in a passenger cabin, along with Group Captain More, a British aviator, and Colonel Melton of the American Air Corps. All three were bound for the Senior Officers' Camp in Formosa. With Varley cut off completely from his men the Australian POWs now had no combatant officer more senior than captain and were dependent for leadership on Arthur Sumner. John Chalmers and Rowley were the only Australian medical officers on board.

There were 716 other Australians on the *Rakuyo Maru*, the fittest of Kumis 35–39, and 599 British. On the *Kachidoki Maru*, the other passenger cargo ship that had been loaded in Singapore and had set out with the convoy, there were, Rowley had heard, about 1,000 British POWs.

Many of the men Rowley had come to know well on the line, his medical orderlies and hospital helpers, had been left behind. He had no way of knowing what had happened to Jim Armstrong, John Shaw and Don Kerr after he left Tamarkan. The youngsters Roy Whitecross and his friend Maurice Barkley were both in Kumi 41 which remained in Singapore, presumably to join a later convoy. But he still had Len Coon as medical orderly. The nimble-fingered Jackie McKone was acting as his batman as McKone's officer, John Ross, had also been left behind. He adopted Rowley as his officer.

Rowley realised when he had time to think about it that he was lonely without Jim and John Shaw. In splitting up even the kumis of the Japan party the Japs had finally achieved disintegration of the group identity that had been so important to survival on the line. Small knots of friends had also been broken up, leaving a general sensation of insecurity. So there was a strong urge for comradeship among the Australians aboard the *Rakuyo Maru*. Rowley found that Arthur Sumner was a good team-mate. Together they had worked on a system of evacuation in the event that the ship was torpedoed.

Leaving the fetid hold Rowley took up his station just behind a mast where he had established his RAP for Kumis 35, 36 and 37. John Chalmers was looking after the other two kumis. Some of the sick, cyclic malaria cases and some lingering dysentery, were already lined up for sick parade; others were queued up for their turn at the latrines—mere wooden frames slung over the side of the ship by ropes and of which there were three on the port side for the Australians and three on the starboard for the British.

'Trying to go when you are hanging out over the side of a ship is a bit nerve-racking,' McKone observed. 'But at least there's a breeze and the view's good, Doc.'

The view was of the great, flat circle of ocean from which the tropic glare of the sun was reflected with eye-stinging brilliance; featureless except for the strung-out convoy. There had been eleven ships when the convoy left Singapore: the two passenger-cargo vessels, two tankers and two smaller cargo ships, with five escorting destroyers or frigates. The *Kachidoki Maru* was of about 10,000 tons, and the *Rakuyo Maru* of 8,000 tons. Off The Philippines three more cargo ships and three escorts had joined the convoy. One of the small escorts from Singapore, limping with engine trouble, had left it to put in to Manila.

Rowley stared thoughtfully at the smooth, blue glare of the ocean, as he had done on each of the six days since the convoy had left Singapore. Where beneath the surface were the submarines lurking? It was not a question of were there submarines: even the POWs in Singapore knew that the American wolf packs were causing havoc among Japanese shipping in the South China Sea; it was simply a question of where were they? Would their blockade, by chance, miss this convoy? Or, if and when they struck, would their torpedoes miss the *Rakuyo Maru*?

'I'm getting used to the mixture of salt water and fresh,' McKone volunteered. 'Reckon I can stand fifty-fifty now.'

Len Coon, who had taken his place on Rowley's right with the sick roll, chimed in, 'If we have to I'd get by drinking sea water straight for a few days. Although it still doesn't taste so good.'

'But more attractive than having to drink urine, eh?' Rowley said.

Coon and McKone, he knew, had been sharing his thoughts about the submarines as they stared at the sea. If they were adrift on rafts and debris for several days, supposing the submarines weren't able to pick them up right away, thirst would be their greatest problem. There were stories of shipwrecked people drinking their own urine and, of course, it was generally believed from tales of the sailing ship days that the shipwrecked who drank sea water went raving mad and died horribly.

Rowley's experience with the use of salt and salt water for cooking led him to believe that the body could tolerate it providing the kidneys, stomach and palate were accustomed to it. As a result of his observations at Changi, and Tavoy and Hlepauk he had concluded that the body could be trained to the intake of salt water provided it was not already so dehydrated that concentration of salt in the bloodstream robbed the tissues of water. The gruesome stories of shipwreck victims going crazy after drinking salt water, he believed, were explained in that they were already so dehydrated even the brain cells were rapidly robbed of water. The important thing was that the body should not be allowed to become dehydrated before the intake of salt water commenced.

In fact Rowley had commenced experimenting from the time the Japan Party had been formed. On Jeep Island and now on the ship he had trained himself and encouraged his men to drink a mixture of salt and fresh water and had reached the stage where he could manage to tolerate almost straight salt water without ill effects.

If they weren't picked up and had to try to sail whatever craft they had to the nearest land, it was possible they might even last ten days by mixing fresh and salt water, provided there was rain.

But on a clear day with a clear sea that line of thought was not profitable. Although the thought of submarine attack was ever present, it was best to keep it at the back of the mind.

'It's time we had some luck. We'll probably get through,' Rowley said. 'Now . . . we have a sick parade to do.'

There wasn't very much he could do for the men on sick parade other than listen to their symptoms, list them in the medical records and give some words of advice and encouragement. He had nothing else to give now. He was a doctor practising with the barest minimum of medical supplies.

Sick parade over, he went to the rail and watched the strung-out convoy. *Rakuyo Maru* was in the centre of the starboard flank so that her starboard side was open to the menaces that might lurk in the deep. He could see *Kachidoki Maru* to the port and behind, with the two tankers in the centre and a smaller cargo vessel leading, on the horizon. The destroyers were low smudges on the sea off the starboard flank.

Arthur Sumner joined him. 'I wonder where they are.'

'We all wonder. How is your evacuation program?'

'I run the men through it in groups every day. Orderly evacuation of the hold; toss over everything that'll float, the rafts if the Japs don't get them all first, hatch covers, the latrines. We get bayonets off the guards to hack them free. You keep your sick on deck and get them over the side first. But Rowley, if she goes down quickly we'll be bound to lose a lot of men in the hold. That stair isn't big enough to get them all out.'

'Maybe the subs will miss us,' Rowley said. 'Let's get below for a while. It's stifling in the hold but it's scorching on deck, with no shade.'

'The Japs travel first class. Dining saloon, iced water . . . and even their popsies. Did you see the little girl with Yamada on the Bridge? I wonder what she thinks about the submarines,' Sumner said.

'She's busy doing her duty for the glory of Nippon.' Rowley grinned. 'I hope they trained her to swim as well as other things. Cheer up. Every day afloat is one nearer Japan and the end of the war. I'll drink champagne at Romano's with you yet.'

Some time in the early hours of 12 September the submarines attacked. Rowley, wrapped in a thin blanket, was stretched out on the deck in his RAP position just behind the mast. He came awake aware of an explosion, but at first could not decide what the racket was about. The Japs were running and shouting and firing off their deck guns. What their target could be he could not imagine. Then there was another explosion and a great ball of fire appeared on the sea ahead and to starboard. It was there, then suddenly it was gone, sucked down into the sea.

'That was a destroyer,' Jackie McKone said at his elbow. 'Jeez, she went quick.'

Coon was also there. 'Orders, sir?'

All over the for'ard deck Rowley saw that men were stirring on their roosts. The demise of the destroyer was cheered and there was a chorus of boos as the muffled explosions of the depth charges started. There was no panic among the prisoners. Japs were running and shouting but he couldn't make out if that was panic or routine procedure in case of submarine attack.

'Get around among the deck sick. Warn them to be ready to leave but not to go over before they have to. McKone, let's check the hold.'

He found that the stairway to the hold was spewing up a steady stream of men. Someone touched him and said, 'Captain Sumner is down there, doc.'

He called down the stairway and Arthur's voice answered. 'Everything under control. Don't come down. You'll block the way up.'

'D'you think, with the Jap confusion, you can get some bayonets?' he asked McKone. The little man nodded and slipped away.

Rowley decided there was nothing he could do at the moment except stay at his RAP position where he could be found when needed. He checked that his waterproof wallet was safely in his pocket.

The battle with the unseen submarines was a pyrotechnic display that reminded him of fireworks on Sydney Harbour. When the tanker in the centre of the convoy was hit it became a mass of flame on the sea that lit the whole scene as if daylight had come, although it was not yet quite dawn. He could make out the *Kachidoki Maru* to port. She was picking up speed and veering westward, making a run for it. The engines of the *Rakuyo Maru* took on a different note and he felt the added vibrations in the deck. They were moving faster, too. A destroyer turned and began slicing through the water towards them. There was an explosion, fire and smoke and it was gone.

'They're coming . . . on the starboard . . . two trails,' someone yelled. He thought the voice was British.

There were two dull thuds, one following quickly after the other and he felt vibrations coming up through the deck, seemingly right under his feet. The same British voice called that the hit was amidships, below waterline. Rowley waited for an explosion but there was none. He wondered if the torpedoes could possibly have been duds but felt the ship slowing and decided that the hull had been holed.

'We've got bayonets, sir,' McKone said.

Rowley took the blade that was offered to him and saw that there was a group of men with McKone.

'Right, we get the latrines away first,' Rowley said.

He found that he was giving orders and doing everything that he had rehearsed in his mind repeatedly, but he felt curiously numb. Undoubtedly there were dramatic scenes on the ship and around it that he should have been noting for future descriptions but he found that he saw nothing but the action that immediately concerned him and his group.

The three latrines were quickly away, then he was working with others to get sections of the hatch covers up and over the side. He heard himself cautioning men not to jump while timbers were being thrown over but people were jumping anyway, Japanese as well as prisoners.

The Japanese were neither helping nor interfering with the POWs. They were concentrating on their own evacuation, getting lifeboats filled and away and simply ignoring the others. Rowley saw the captain leave, using a small boat that was kept on one wing of the bridge. His steward was with him and, astonishingly, one other man whom Rowley recognised as a POW. The captain did not seem to be bothering about his presence. Rowley was busy organising his sick so that they wouldn't jump while timbers were still going over.

'Right, sport, it's your turn. Over you go.'

He leaned on the rail and was vividly impressed by the phenomenon of the bodies seeming to fall in a long, slow backwards curve before they hit the water. Of course they weren't falling in a curved line at all but were going straight down. It was the forward motion of the ship that made them seem to curve. The *Rakuyo Maru* was still moving, although more slowly and the throb of her engines had ceased.

Suddenly it was dawn and the ship, wallowing, was very quiet; no gunfire, no shouting. But there were still people on board. Occasional figures moved on the decks intent on purposes of their own.

'Well, McKone, she doesn't seem to be in any hurry to sink. I think we might stay aboard for a while and look around,' Rowley said. 'There aren't any rafts or boats left up here. We could take a trip aft.'

On the way they searched through the Japanese quarters for water-bottles. Rowley had discovered that he had lost his and also his stethoscope with the summary of his war diary in its tubing. Despite the dangers of the moment he felt a pang for the loss of his diary which was his last personal, physical contact with 'Anderson' Force.

Reaching the stern they found that the Japanese had not taken the starboard and port aft lifeboats. There were also some stacked rafts. A group of men, including Arthur Sumner and Lieutenants Les Stupart, Laurie Phillips, Jim Burke and Campbell-Smith, were working at the port boat but it wouldn't shift. The davits were rusted. After another escort was hit and exploded they set about launching a raft and left the ship with their group of men, but Rowley declined to go with them. He was shepherding a group of his sick who had not jumped and decided that they would have more chance in a properly equipped lifeboat. He went to investigate the starboard lifeboat.

He was joined by Frank McGovern, a sailor from HMAS *Perth*, and several other men. McGovern took over direction of the attempt to launch the boat while Rowley checked what supplies were aboard. There was a small cask of water that was stale and a tin of hard biscuits. Several men, including McKone, went off to hunt for supplies while the *Perth* men started to re-rig the davit ropes which they found had been cut.

When run through the blocks the ropes were too short and the boat was dangling some twenty feet above the water. Rowley, who was still carrying the bayonet McKone had given him, gave it to McGovern who sawed at the stern rope while Russ Savage used a knife on the bow rope. The boat crashed into the water and bobbed right side up, but when the launching group reached it they found the fall had sprung a plank and the boat was leaking. A plug of blue cloth was not completely effective and the party, including Rowley's sick, took turns at bailing.

McKone dropped down into the boat slung with water canisters—he had found and filled five of them—and clutching a tin of biscuits. McGovern and his men from the *Perth* organised the rowing and the leaky boat pulled away from the side of the *Rakuyo Maru*.

Once away from the ship they began to pull in survivors who were clinging to debris or swimming. Rowley saw Jimmy Mullens, a Manly boy who was a great surfer, swimming from one piece of debris to another, assisting survivors and towing them to the lifeboat. He was having a ball, revelling in freedom and the water, frisking like a happy puppy. He handed several weary men into the lifeboat before coming aboard himself.

When there were some sixty survivors in and around the boat Frank McGovern suggested it was too risky, with a leaking craft, to try to help any others. As it was, those aboard took turns in going over the side and hanging onto the gunwale while the remaining crew bailed the boat. Rowley saw that one of the Japanese 'hors de combat' was aboard, sitting very still and sensibly quiet. Someone had found her abandoned and crying on board and escorted her into the lifeboat.

During the day they drifted in the vicinity of the *Rakuyo Maru*, watching for some form of rescue. They saw men reboarding and leaving the stricken vessel. Scattered over a wide area of sea were lifeboats full of Japanese, rafts piled with men, some Japs, some white, and men in the water swimming or clinging to debris.

One Japanese lifeboat came close to the prisoners' boat and the occupants demanded the biscuits, the water and their young woman. As they were armed it seemed safer to hand over the supplies which they would sadly miss, and the girl with whom they were happy to part, rather than risk losing their boat entirely.

Later they saw a raft return to the *Rakuyo Maru* and figures go aboard the stern, setting to work on the remaining lifeboat. After some time they succeeded in getting it launched.

No submarines appeared but late in the afternoon the remaining Japanese destroyers returned and began collecting their people from the lifeboats.

Obviously they had no room to spare for prisoners and were doing a fast rescue job for fear of lurking submarines.

As the destroyers completed their rescue mission and raced away, soon to disappear over the horizon, Rowley and the *Perth* sailors began a boat recovery operation, bringing their boat close to one of the empty ones, putting a few men aboard with instructions to pick up swimmers until they had a safe load and then move well away from the rafts again. They discarded the leaking boat which they later saw taken over by some men on a sinking raft. The operation continued until ten Japanese lifeboats had been recovered. The little flotilla of eleven kept close together and at dusk was joined by the last boat to have been taken off the *Rakuyo Maru*. Len Coon was aboard, as well as Campbell-Smith, who was in command. No one knew where the other lieutenants had gone. Rowley asked for news of Arthur Sumner and John Chalmers but no one had seen them. They shouted, however, that they had seen Varley, the Brit and the Yank in a boat.

Before dark the *Rakuyo Maru*, which had been settling slowly but steadily, passed out of the scene, listing and quietly slipping down stern first in the darkening ocean.

The wind had been rising and the sea becoming choppier for some time. Because of the pitching and rolling of the lifeboat the men who were in the water had difficulty in keeping handholds. Rowley saw the fingers of a hand, white and wrinkled with immersion in salt water, begin to open slowly.

'Hang on there, mate!' Rowley lunged for the hand, had a momentary glimpse of an upturned white face, then hand and face were gone. He stared at the choppy water. With no experience of light craft Rowley realised suddenly and without any shadow of doubt that he was seasick. He yelled 'Look out!' then vomited. Having had nothing to eat for nearly twenty-four hours the vomiting was a ghastly, painful dry retching that racked his whole body.

Someone was holding him and pushing the metal neck of a water bottle between his teeth. It was McKone. Rowley took a couple of gulps before the darkness cleared from his mind and he found the strength to wonder about the water bottle.

'Where did you get that?'

'I happened to be sitting on it when the Japs arrived,' McKone grinned. 'Take some more. You need something to throw up.'

Rowley turned his head away from the proffered canister, hung over the side and vomited again.

'Don't waste that water,' he gasped. 'We'll need it.'

He collapsed into the bottom of the boat too weak to take any interest in its organisation for the time being. Vaguely he was aware that a boat had come alongside them and someone was having a shouted conversation with Vic Duncan, a petty officer from the *Perth*. It seemed that their contingent had decided to sail to The Philippines and were suggesting that the other boats join them. Duncan replied that they intended to sail westward for China.

'Is Doctor Richards with you?' Len Coon's voice called.

'Yes, he's here somewhere,' Duncan answered. 'Hey, Captain Richards!'

Rowley tried to answer but found he could neither lift his head nor control his throat. Another spasm of dry retching gripped him.

'Tell him Corporal Coon said "Cheerio". I'll look him up in Sydney,' Coon sang out.

Rowley tried to croak an answer but no sound came. He let his head roll on the water-slopped decking, ashamed that he was too weak to say goodbye to one of his orderlies.

Later, the four remaining boats were tied together in line, stem to stern, so that they would not lose each other in the darkness and Duncan detailed two men to stay at the oars in each so that they would not be bumping together in the choppy sea. Duncan and Jimmy Mullens set the sail that they had rigged during the afternoon.

'We'll make some way with this breeze,' Duncan said, 'but the bloody compass is no good. The lighting apparatus doesn't work.'

Rowley then discovered that although the sailors from the *Perth* could recognise some of the northern stars, none of them was experienced in navigation. As his training in the Survey Regiment had included orientation and he could use the stars to maintain a bearing, the seasick landlubber doctor was appointed the navigator. For hours he sat by the man at the tiller, keeping the string of boats heading west. At last, at about 0300 hours, he became sleepy. He handed his wristwatch over to a lad at the tiller telling him how to keep the bearing by the stars and to wake him in an hour. Then he found space on the crowded deck to curl up and fell into exhausted sleep.

Before dawn the lad, distressed, awakened him. 'Captain Richards, I've lost your watch sir. I had my arm over the side. The catch must have come loose. It fell off.'

'Can't be helped,' Rowley told him. But he felt sad about the watch. It was Jim Armstrong's. Jim had given it to him in the jungle when his own watch had stopped.

After daybreak there was a redistribution of people among the four boats and they were cast off from each other. Travelling tied together they had bumped continually and had made little progress. Among the fit with Rowley were Jim Mullens, Jackie McKone, Kitch Loughnan, Lindsay Savage, Ian McDermott and Tom Moxham who they'd pulled out of the water. He was now, he joked, officer in charge of a floating RAP.

The *Perth* men and Jim Mullens took charge of the sails. Frank McGovern was appointed quartermaster and with Rowley checked the supplies. They had some hard biscuits, one five-gallon cask of fresh water, the canister that McKone had retained and some cigarettes that men had scrounged from the Japanese quarters before leaving the *Rakuyo Maru*. Also, they had a couple of tin buckets that McKone had collected and concealed from the Japs.

Calculating that they would take at least twelve days to reach the China coast, Rowley decided on a ration of a quarter of a pint of water, two biscuits and two cigarettes morning and night for each man. The day was showery and they began collecting rainwater in a small spare sail and filling the tin buckets. By adding sea water they were able to double their supply and set aside the cask for emergencies. Undoubtedly the squalls would not last and there would by many rainless days before they reached land.

All day on 13 September they sailed westward before the easterly breeze. Once the rations had been settled and all knives collected by one of the *Perth* petty officers as a precaution against any thirst- or heat-crazed men running amok, Rowley settled down to navigate and take care of the sick. Jimmy Mullens and the *Perth* sailors had the handling of the boat.

The three boats that had been tied to theirs during the night separated and were sailing in the same direction, spread out a few hundred yards apart. On their northern horizon, which was limited because they were at sea level, they could see a convoy of seven lifeboats on a parallel course to their own. The boat containing Campbell-Smith and Len Coon had apparently left the area during the night on its attempt to reach The Philippines.

Morale was good. They were free men for the first time in two and a half years. It was possible that they would be picked up by American submarines from the pack that hit the convoy and would put their feet on dry land in the United States, still free men. If they weren't picked up, with luck they might make China. Even if the future there was uncertain it could be no more uncertain that it had been at any time since capitulation.

Men were so sure of being picked up that they even began speculating on living conditions aboard the submarines, supposing that they would be cramped and closed in but the discomfort would be worthwhile because they would be free and with friends. When that subject was exhausted and no submarines had appeared Rowley started reminiscing about the Malayan Campaign. It was as well to keep the men talking, keep their minds off their hazardous situation in a lifeboat on the South China Sea—and of the fate of the mass of men they had had to leave behind, crowded on rafts and hatches, clinging to debris, or just swimming.

They relived the Malayan Campaign, discussing the contribution of the Malays, the British Army and the Australian Army, the efficiency of army training generally, the whole debacle of the fall of Singapore. After capitulation had General Gordon Bennett been right to attempt escape or should he have stayed? That was a point that had been argued around camp fires from Changi to Tamarkan and opinions were still divided. The *Perth* men and the British survivors in the boat thought that Gordon Bennett had left before capitulation but Rowley was able to correct that misapprehension, describing how he had actually been in headquarters at Tanglin when the General announced the capitulation.

They remembered the march into Changi, the bug-infested beds, the Changi University, the building of Tavoy aerodrome. As they talked the squalls cleared and the sun moved over the mast tip. It was hot. Rowley and Kitch Loughnan made an awning from the spare sail in an attempt to shade the weakest of the sick, who lay on the decking. Rowley measured out an extra ration of the salt-freshwater mix for the sick and McGovern, the quartermaster, handed around a cigarette apiece.

Mills, who had had his jaw broken on the railway when a Korean guard bashed him with a rifle butt, began a yarn about tricking the guards into short measuring the number of yards of earth the labourers were to move in a day, assisted in the details by his cobber Pat Major, a fresh-faced, cheerful young AB, also from the *Perth*, who was taking his turn at the tiller. That in turn prodded the memories of the two Queenslanders, Kitch Loughnan and Lindsay Savage, who chimed in with stories of guarding and moving the miserable cattle that had been the meat ration for their group. Both were cattle men, Lindsay from Kingaroy where he bred Red Polls and Aberdeen Angus and Kitch from Mitchell, an area favouring horned Herefords and Shorthorns.

Kitch, a gangling six foot three, was jackknifed on a small area of decking with his long legs drawn up, bony knees under his chin. He took a last, appreciative draw on his cigarette, pinched it out and tucked the bumper over his ear, then settled down to his favourite ongoing argument with Lindsay over the merits of horned against poll cattle. Rowley was pleased that the talk went on in a steady stream, keeping the minds of his men off immediate and future dangers and from turning to the fate of the men that had been left behind in the water.

It was stinking hot.

Jimmy Mullens, tending sail, called that the convoy of seven boats that had been sailing parallel with them all day had turned about and was heading eastward into the wind. 'They must have decided for The Philippines after all,' he said. 'The Brigadier will be in charge of that lot, won't he, sir?'

'Most likely,' somebody affirmed.

Conversation dwindled and died as they watched until the seven boats had disappeared from their view. Now only four lifeboats sailed the otherwise empty circle of ocean.

The sun dipped down to the western horizon and the breeze dropped entirely. The sail was limp against the mast.

'We can row. We can row all night while it's cool,' Pat Major suggested. Mills agreed with him and Jimmy Mullens, always irrepressible although his strength was not as great as he thought it was, supported them. Kitch and Lindsay said they were willing to take a turn at the oars but Rowley vetoed the idea.

'No way. We can't row all the way to China. We'll need all our strength for sailing when we have a breeze. Another thing . . . it's best to stay in sight of the other boats. The subs are more likely to see a group of boats than a lone one. We'll ride out the night here.'

Mills broke out the rations—a biscuit, a quarter pint of water and a cigarette each. They sipped and chewed slowly, scheming to defeat hunger by spinning out the length of the meal. Still it was finished all too soon. The sun had gone and the darkness of the tropics came quickly. The mast tip described a circle against the northern constellations. There was a brief shower and McKone busied himself pouring off the water that had collected on their makeshift awning into one of the tin buckets.

'At least it's cooler now,' someone said. 'Just as well the bloody sun doesn't stay up for twenty-four hours.'

Later the night seemed quite chilly in contrast to the heat of the day. There was a heavy sea dew and Rowley found that he shivered a little in damp clothes. The changes of temperature, he thought, would not be good for those men still in a malaria cycle. Malarial delirium in an open boat might be hard to handle. He heard restless men shifting position on the decking and occasionally someone groaned a little. The lapping of the ocean at the sides of their craft was continuous and monotonous. There was no other sound. He felt very aware of the dreadful loneliness of the sea. Would the submarines be able to find them?

The sun came up but did not bring a breeze with it. The other boats were still in sight. Mills handed out the rations. The quarter pint of fresh water seemed a pitifully small amount to get them through a day of blazing sun, even when a measure of sea water was added according to how much each man felt he could tolerate.

'I prefer to drink sea water,' Mullens said in an affected falsetto. 'The flavour is so much better. Wait till I tell them that at home!'

Still there was no breeze. The men bailed in shifts of two.

'Sir, can't we get on and row?' Major asked. 'I don't feel right doing nothing.'

'No, the other boats are drifting. We'll stay in sight,' Rowley told him.

'We can sing,' Kitch volunteered. He started 'Waltzing Matilda', presumably with the idea that it was one song everyone would know. Others began to join in.

'Cut the noise,' Mills called suddenly. 'I can hear something. Damn it, it's gunfire . . . light, rapid-firing guns.'

Everyone listened. At first it seemed that Mills was imagining things. Then others heard the distant rattle.

'Would that be submarines?' Mullens wondered aloud.

But American submarines did not find them. Jap frigates did.

They had been speculating for about an hour about the meaning of the explosions which seemed to have sounded to the north—was it more ships, more torpedoing?—when a frigate appeared on the horizon, followed by two more. They seemed to be making a sweep across the ocean.

'Looks like they're looking for something,' Kitch said.

One of the frigates swung around and came knifing directly towards the widely spread lifeboats.

'Us!' Mills yelled. 'The bloody bastards have come back to finish us off.'

Rowley found that Mullens had grasped his hand and was shaking it. Throughout the boat men were shaking hands.

'Well, that's that. It was nice knowing you blokes.'

'At least we beat the bloody jungle.'

'No more spoons and plates to be sterilised, Doc.'

'What if a bloke is wounded but not killed? I'd like to be put out of me misery.'

'Me too. Will any survivors please shove the wounded on their way, nice and easy.'

There was general murmured agreement that any one who might survive the gunning should dispatch wounded who were unable to take care of themselves.

They watched as the minutes passed and the frigate approached the first of the lifeboats to the north. The frigate slowed until she seemed to be carried only by her own momentum, then picked up speed again. The lifeboat was empty. Yet there had been no sound of gunfire.

'Well I'll be . . . They've picked the boys up,' Kitch gasped wonderingly.

'What's the trap? Are they going to question them before finishing them off?'

The frigate bore down on a second boat, slowed, then picked up speed. It repeated the manoeuvre with the third boat and was now heading towards them.

'They are picking us up,' Rowley said. 'Come on. Get the sick blokes on their feet. They won't give us much time.'

'They have scramble nets out. We grab and climb,' Mills yelled.

Rowley had a confused impression of the moving side of the frigate, Japanese faces looking down and something that appeared to be a rope ladder or a net swinging towards him. He steadied the sick man he was supporting, gave him a shove up and saw him grasp the ladder thing. Then his own hands gripped rough rope and he was climbing.

They jumped, stumbled and fell on to the deck of the frigate. Rowley picked himself up, looked over the side and saw the empty lifeboat sliding away into the frigate's wake. No one had been left behind.

Before Rowley could properly comprehend the rescue he and his companions were being hustled into the bows of the frigate by the Japanese crew. Amazingly, survivors from the nearer boat were huddled there. Major and Mills were actually hugging Petty Officer Johnson. Rowley saw the brothers Ernie and Jack Wade and Ross Dunbar of the 2/15th. Lindsay Savage found his young cousin Russell had been picked up.

A Japanese officer made signs for them to move as far up into the bows as they could, sit down and keep quiet. He pointed at the deck guns, whose crews were at battle stations, then at sailors armed with machine-guns. Obviously the crew was alert for action and the survivors would be gunned down if they got in the way.

With the castaways huddled for'ard the frigate headed on a generally westward course, zig-zagging continually.

'They're manoeuvering to avoid torpedoes,' Mills said. 'Must think the Yanks are still around. They have to hope they don't zig when they should have zagged.'

'I hope so too,' Rowley said. 'After seeing two escorts hit I don't fancy being aboard when one cops a torpedo.'

Rowley stared at the Japanese navy crew, a better lot in smartness of appearance and discipline than the army or merchant ship fellows the prisoners had been used to. They were nervously tense, officers scanning the sea with binoculars. He presumed that this and the other two frigates darting back and forth in the distance were the survivors of the convoy escort . . . the ones that had picked up the Japanese from the boats. They must have off-loaded the Japs somewhere and come back on a mission to rescue POWs, even though they must have realised the subs were probably still lurking in the area. Who had ordered the rescue mission? High Command? Local Command? Or the frigate's commander? What a truly amazing, unpredictable people.

Rowley sat with his chin on his knees, feeling the bite of the tropic sun on his fair skin. Pat Major, he noticed, was badly sunburned, his fresh face blistering, and others were suffering varying degrees of sunburn. They talked quietly among themselves, asking about mates who were not on board. From the snatches of talk Rowley began to form an idea of the scene where the rafts and swimmers were massed . . . prisoners taking their revenge on Japanese in the water for the sufferings they had endured on the line. Someone had seen a bloke get his hands on the throat of a Jap and the two had gone under together. Someone else had seen a man on a portion of a hatch cover cracking the heads of Japs in the water with a chunk of wood and said, 'I saw Dodger kicking along in the water like a kid with a float board, steering part of a latrine platform, and he had the damned thing loaded with bottles of grog and fruit he must have pinched from the dining saloon. Trust that old Dodger to look after himself, I thought, but when I looked again he had gone. Must have been a cramp that got him.'

'Wonder how long the blokes in the water and on rafts would last? Things would be pretty grim for them.'

Vic Duncan edged his way up beside Rowley.

'Sir, the other seven boats . . . the ones that went east . . .'

'Yes, I'm wondering about them,' Rowley told him. 'I'll ask.'

The frigate crew seemed to have relaxed somewhat, although still at their battle stations, so he went to the edge of the for'ard deck and asked in his passable Japanese to speak to an officer.

The for'ard gun crew seemed amazed that he could speak their language but obviously understood. One called to a petty officer and there was an exchange of remarks too rapid for Rowley to follow. Then the officer went away and shortly returned with a youthful looking two-ringer who was, Rowley assumed, the captain.

Rowley bowed, observing necessary Japanese ritual.

'Richards *Tai* [Captain] Australian doctor,' he introduced himself.

'Nagahashi *Chui* [Lieutenant],' the captain responded.

'I am the only Australian officer here,' Rowley explained. 'Other officers and men were on seven life boats. Can you tell me what happened to them?'

The captain listened attentively, pointed to the other two frigates, then waved for Rowley to go back to his place. The meaning of his gesture was unclear and Rowley frowned, remembering the gunfire they had heard in the early morning.

For the present the fate of Brigadier Varley, the British and American senior officers, Captain Sumner and Lieutenants Stupart, Burke and Campbell-Smith was uncertain. And what about Major John Chalmers?

'Did anyone see Major Chalmers?' Rowley called out.

'I saw him on a raft just before a lifeboat picked me out of the water,' someone answered. 'There were a few blokes on the raft and the boat was already loaded. We didn't go in.'

In the late afternoon some of the frigate crew brought around a ration of a third of a pint of water and half a ship's biscuit per man. All night the survivors huddled on the frigate's deck. At about ten o'clock on the morning of 15 September they sighted land which they thought might be the China coast. Later they were told that it was the island of Hainan.

The three frigates slid into the harbour of Hainan To, which was crowded with shipping: small cargo vessels, some navy frigates, an oil tanker and a huge vessel of unusual design anchored in mid-harbour, which proved to be a whaling mother ship. The frigate slid alongside the oil tanker and her crew started herding the prisoners on to that vessel. The other two frigates were now close enough for Rowley to see that there were no prisoners on their decks. Again he approached the captain and asked what had happened to the men in the seven other lifeboats. Again the young captain pointed at the other frigates but said nothing.

Rowley could still not be sure what he meant. The seven crowded lifeboards had sailed over the horizon and vanished. Their fate could be presumed but was never established. As Rowley left the deck of the frigate for the oil tanker he realised with a dull flicker of outrage, for repeated disasters numb the senses, that he was quite possibly the only Australian officer to survive the sinking of the *Rakuyo Maru*.

11

Picking up the Pieces

'Oh God!'

Nothing that Rowley Richards had seen on the Burma railway, not even the death huts, had prepared him for the scene on the deck of the oil tanker in the harbour of Hainan To.

There was an expanse of shimmering metal deck and some sort of superstructure that cast a shade, and in the space beneath this superstructure men were huddled or sprawled like straw-stuffed dummies in every little patch of shadow. There seemed to be hundreds of them, naked, blackened travesties of humanity, a Goyaesque picture sharply delineated in white light and dark shadow and made more terrible by the steady vibration of deep-tortured groans and weak moaning.

Rowley's legs carried him forward in an involuntary rush into the shadow. Then the gut-wrenching stench of oil and hot, raw, burned flesh hit him and with it an engulfing wave of sorrow that bodies could be so abused and suffering. He retreated and heaved, not caring that he was going to spew on the metal deck, but nothing came from his empty stomach.

When he lifted his head he saw that the survivors off the frigate were grouped behind him in petrified stillness, horror on frozen faces and in wide eyes.

'They're ours . . . white people,' he gasped.

The group slowly began to move and talk.

'What do we do?' Jackie McKone asked.

'Find out what happened, what the injuries are. McKone, Duncan, take some men and try to make these people easier . . . but be careful about breaks, internal injuries. Report to me for directions on treatment.'

Rowley hoped that he sounded confident and efficient for he was the only officer in the *Rakuyo Maru* group and had to provide leadership. To keep your act together and do something in an impossible situation was, after all, what the rigorous training of the military machine was all about. Give directions for treatment! What treatment? What did he have to give?

'When there's nothing else to do you look cheerful and give them hope. Amazing how people pull through on hope.' They were the words of Arthur Lewis, Rowley's uncle talking from years ago in Kyogle Hospital. 'Never let the patient see you can't cope.'

With Jimmy Mullens assisting him Rowley began to move among the mass of bodies, willing himself to observe, assess, diagnose, attempt to devise treatment. The black bodies were thickly covered with oil. Rowley assessed first- and second-degree burns, both oil burn and sunburn. Gently he moved the hands that a man held over his eyes and the man cried out in pain as sunlight penetrated between swollen, reddened lids. There was sun and oil burn damage of conjunctiva and cornea in many cases. There were fractured limbs.

Even the British officers from the *Kachidoki Maru* were so far gone in suffering that they were unable to take care of their men or themselves. Despite cracked, swollen lips, puffy lids and reddened face, one officer slumped against a bulkhead was recognisable: Jim Roulston, Royal Army Medical Corps, whom Rowley had encountered on the line.

'Glad to see you, Richards. We need doctors', Roulston whispered weakly.

''Fraid I'm the only one. You were on the *Kachidoki Maru*?'

'The Yanks got her just outside Hainan. We're all that's left, I think.'

Roulston tried to sit up but fell back. He had fainted.

'They've hit bottom,' Rowley thought. 'We've got to do something for them—anything. Just so long as we are doing something.'

He found that his stomach was knotting up again and hurried for the open deck, gulping in air that was redolent with the smells of a fishing harbour but blessedly free of the smell of oil and burnt flesh. Afterwards he had no clear recollection of the rest of that day, only fragmentary impressions of attempting to obtain buckets and rope to draw up sea water for washing away the flesh-biting oil, of attentive Japanese faces that seemed vaguely sympathetic, and of being given the buckets but nothing else. Heads were shaken when he asked for splints, bandages, lotions for the burns. He thought they either did not understand or had no supplies themselves. He heard men screaming out as their bodies were washed with sea water that stung raw flesh. But the oil had to be cleaned away. Despite the excruciating pain, even eyes had to be bathed with salt water. At least it was cleansing.

Five or ten minutes at a time was all he could manage among the mass of the injured before he had to seek fresh air and heave.

When the sun went down and the air was cooler the injured seemed better able to tolerate the bathing. For the fracture cases there was little to do except place them in positions which would best protect the limbs. Later, rags of clothing were used to improvise slings and bandages.

Then it was dark and the exhausted survivors from the *Rakuyo Maru* slept on the deck, the moaning of the survivors from the *Kachidoki Maru* running through their nightmares.

On the next day the British, Australian and Japanese survivors from the convoy were transferred to the whaling mother ship. The blinded had to be led, those with fractures carried. They were stowed in the cavernous, echoing, blue-lighted metal belly of the ship, an enormous hold that seemed to extend for its full length. The prisoners were herded into a bow section where at least each man was allowed enough space to lie outstretched. That afternoon they sailed in convoy for Japan.

The Japanese survivors from the convoy were accommodated on deck. Both they and the prisoners were fed from the same kitchen—rice, barley and a little meat in vegetable stew—and it appeared that the Japanese had no medical supplies even for their own injured.

Six British officers had survived the sinking of the *Kachidoki Maru*: Captains Wilkie, Peirce, Pote-Hunt, Barrett, Jim Roulston and another officer who was badly injured and only semi-conscious. Rowley was the only Australian officer survivor of the convoy as far as he knew.

The *Kachidoki Maru* had almost succeeded in her run for cover but had been torpedoed late on the night of 12 September, about twenty-four hours after the convoy was first hit. The transport went down within twenty minutes and many men were trapped in the hold, having delayed evacuation apparently on the assumption that the ship had not been badly holed.

'A lot of us were caught in a ghastly oil slick,' Pote-Hunt told him, 'It was thick on the water. You could see the greasy shine of it and it burned like the dickens. Fellows couldn't help swallowing some.'

Bit by bit Rowley heard of the *Kachidoki Maru*'s swift stern-first dive under the ocean. Abandoning of the ship by the prisoners seemed to have been unorganised and too slow and the officers were of the opinion that about 300 prisoners had been trapped in the hold.

The experiences of the British had been similar to those of the *Rakuyo Maru* people, except for the added danger of the huge oil slick and the organised rescue of both Japanese and POWs by the Japanese. Naval ships—a frigate and a couple of corvettes, Peirce and Pote-Hunt thought—had appeared at the scene of the disaster on the following day and began picking people out of the water, the Japanese first and then such prisoners as the ships could accommodate. Then a small trawler had arrived, towing a string of lifeboats that it had picked up. The trawler skipper had loaded his little craft with British before setting the lifeboats adrift for the mass of men still in the water. The Japanese ships had then taken their human cargoes to Hainan To, off-loading them on to the tanker and had gone out again a couple of times to bring in more survivors—an extraordinary act of bravery and compassion since the Japs could be reasonably sure there were still American subs in the area. The prisoners could only speculate on whose orders the rescue mission had been launched.

Once the sick and injured on the whaler had been made as comfortable as possible and the most able had fed the helpless from the rations of rice and weak stew, the small group of five British and one Australian officer made a head count. Of 1,000 British who had been aboard the *Kachidoki Maru* there were 520 survivors. Rowley's group were a mere 136 of the 1,319 Australians and British and one American who had been aboard the *Rakuyo Maru*. The fate of the complement of the other lifeboats remained unknown. As for the hundreds of men in the water, there could be little hope for them unless some were picked up by the submarines, as the Japanese had made only the one rescue trip. That was understandable since the *Rakuyo Maru* had been hit some 200 miles further from Hainan than the *Kachidoki Maru*: probably the Japanese thought any return into that area too dangerous, and also presumed that the men in the water could not have survived so long in any case.

The magnitude of the disaster numbed comprehension, but Rowley found he was more affected by the individual state of the injured aboard the whaling ship than by the knowledge that there were probably only 136 survivors from the *Rakuyo Maru.*

Eight men died of their injuries within a few days. The skipper of the whaling ship, a brisk, neat, naval type, permitted their proper burial at sea. As a body wrapped in a grass mat was dropped over the side Rowley pondered on the harshness of fate that a man should survive the hell of the Burma railway, then die because of the actions of Allied submarines. With anger he remembered that the Japanese had displayed no markings on the ships to indicate that they were carrying prisoners of war.

Although they were fed and had adequate space for each man, confinement in the cavernous hold was an agonising ordeal, not only for battered bodies but for tightly strung nerves. They were kept below except for short periods when they were permitted to use the latrines. Numerous empty forty-four gallon drums were stacked at one side of the hold, only loosely secured, so they rattled constantly. When the ship shuddered and jarred for any reason, the drums banged together with deep reverberations that sent injured men struggling to their feet, calling that they had been torpedoed again.

The ship was moving in company with two other transports, with an escort of three small frigates. When the next submarine attack came Rowley was below. He had just become accustomed to the sound of alarms and excursions on deck, which happened regularly and he had decided were merely the Japanese carrying out gun drill; so it was the sudden dull noise of exploding depth charges, then the sound of gunfire, with vibrations that set the oil drums rolling and resonating once more, that alerted him that they were again under attack.

'Hold it, fellas. Take it easy! We aren't hit yet.'

He began calming his men and heard the British officers doing likewise. There was a panicky babble of speculation but no loss of control or discipline.

A big explosion, followed by the shuddering and slowing of the ship and a seeming altering of the engines, made him think, 'This is it.' As the oil drums resonated to the explosion the din in the hold was terrific. Men hunched their heads into their shoulders and put hands to their ears. 'Oh Christ! Not again!' Jimmy Mullens called out in complaint. 'I don't feel like a swim.'

Then the engines resumed their steady throbbing and the ship moved smoothly. Men who had been up to the latrines dropped down the ladders into the hold, reporting that the explosion had been the demise of one of the escorts. It had run in and intercepted a torpedo meant for the big whaling ship. One of the transports had been sunk, the other damaged. The remaining two ships pulled in to Formosa on 20 September, then proceeded on their way to Japan five days later, accompanied by two escorts. A day or so from Japan there was a second submarine attack, at night. One of the escorts was hit, burst into flames and sank within moments. The other transport was not seen again. Of the convoy that had left Hainan there were now only one escort and the whaling mother ship. The prisoners in the hold were almost light-headed, disaster-happy, talking loudly, laughing when the oil drums clanged. The laughter was a little too high-pitched, the chatter too constant.

The Japanese survivors were going up to the deck in batches, then returning, chattering excitedly and grinning once the coastline of their homeland had been sighted. Such was the euphoria that they took off their life jackets and gave them to the prisoners, saying that they were nearly home and no longer needed them. The prisoners took the jackets and put them on since any piece of equipment might come in handy sometime. A bloke never could tell. The guards posted at the top of the gangway relaxed and joked and took no notice when Rowley and some of the others came up in the hope of being allowed to see the land.

They saw a low coastline with headlands and beaches backed by lines of fir trees.

'Looks like the ocean beach at Manly, sir,' Jimmy Mullens said, his voice low with nostalgia. 'Seems a long time since I went surf-boarding there.'

'Never thought I'd be so pleased to see Japan,' Roulston added. 'I was beginning to feel we were like the ten little nigger boys.' Although the comment was cheerful enough the tone was not.

Rowley leaned on the starboard rail, listening to the excited comments and the talk and laughter of the jubilant Japanese, interested in his first sight of the west coast of Japan. Mullens was right. That beautiful stretch of beach was like Manly. The conifers were tall and dark green, the sand white, the sea blue and innocent. No metal terrors lurked in its depths here.

Then suddenly he saw the wakes of two torpedoes tracing dark lines across the stretch of sea from the direction of the beach, heading straight for the side of their ship. The Japanese watch had seen them too, for the ship was already heeling to port. She seemed to be responding very sluggishly, turning with ponderous, painful slowness. Rowley wondered if she would be able to get around in time. When the torpedoes passed and churned on harmlessly out to sea he exhaled explosively, only then becoming aware that he had been holding his breath and that his hands ached from the force with which he had been gripping the rail.

'That sub must be lying up practically on the beach,' somebody gasped.

Lindsay Savage slapped Jimmy Mullens on the arm. 'Pity the Yanks missed. We could have swum ashore from here. Done a bit of surfing, eh Jimmy?'

Roulston leaned close to Rowley. 'Thank God they missed. This ship has big tanks and I'd say they're full . . . but not with whale oil. I believe she's been converted into a tanker.' His voice was husky.

'The bastards! Carrying prisoners and oil!' Rowley gasped. 'We would be roasted.'

The Japanese, much sobered, took back their life jackets.

Since the American submariners were so daring it was possible that other tin fish were lying off the beaches and the atmosphere was tense as they proceeded up the coast. The prisoners as well as the Japanese survivors tended to remain on deck for as long as they were permitted. There was a marked increase in the need to use the latrines. In the cavernous, dim hold of the whaling ship men now talked in nervous whispers although there was no need to whisper, and flinched when the empty drums rattled. The *Perth* sailors, Rowley noted, were particularly tense. He wondered how many times a man could have a ship torpedoed from beneath him before the psychological damage reached a critical level. Although they were valiantly trying to hold on to their control, some of those boys were already stretched dangerously tight. They could easily snap, go into mad panic and try to abandon ship if an oil drum crashed out of the pile. Once a few started a rush most of the other frightened prisoners would follow and the Japanese would shoot them. The situation was nasty.

He hoped that others had not been as acute in their observations and deductions as Roulston and when the whaling mother ship moved into the harbour of Moji at dusk on 28 September, apparently the lone survivor of the convoy, the arrival was as welcome to the prisoners as to the Japanese.

'It's not a patch on Sydney Harbour . . . nothing is, of course . . . but I'm bloody glad to be here,' Kitch Loughnan said. 'Beautiful, beautiful Moji Harbour.'

The prisoners were unloaded and herded into some buildings that had been stables. There, some clothed in a few rags, many still naked, they spent a miserably uncomfortable night. Rowley heard the injured British officer moaning in uneasy half-sleep. He heard the stumbling departure of one of his own men for the end of the stables where some buckets had been placed for their needs. He thought it might be young Jimmy Mullens. Climbing stiffly to his feet Rowley began walking up and down, a few yards either way, in the aisle between the stalls, waiting for the boy to return. He could not sleep. The floor seemed harder and colder even than that of the hangar at Tavoy and his limbs were stiff and sore. When he was lying down, trying to relax, his body began to shake, so he might as well walk around and have a few words with Jimmy.

The boy was a long time returning and when he came he walked very slowly, dragging his feet. He was breathing heavily, like a tired, elderly man, rather than the bouncing practical joker and baiter of the guards on the railway, or the seemingly tireless swimmer rescuing weaker men after the shipwreck.

'How are things, Jimmy?' Rowley whispered.

'What? . . . You're awake, sir. I'm all right, just a touch of the tom tits.'

Rowley put an arm on his shoulder and felt that he was shivering.

'Walk up and down with me for a bit. It's the only way to keep warm. Jim, you've been great giving a hand with the sick Brits, but you'd better rest a bit. The dysentery keeps coming back whenever a fellow gets tired. And I need you strong young fellows fit. We are going to have to look after our older men.'

'You can count on me, doc. I'll be a hundred per cent in no time. Japan's got to be better than Burma, hasn't it?'

'Of course. We might even find they abide by the Geneva Convention here and treat POWs decently. Food should be better. And there won't be the rain and the heat and filth.'

'There'll be snow this winter. That's what the Jap commander at Tamarkan said . . . that purifying snows cleansed the country and the people. I've never seen snow, sir. Once when I heard it was snowing at Katoomba I started out with a mate who had a motor bike, but the darned bike broke down. Almost seemed as if I wasn't meant to see snow.'

'I've never seen snow, myself, Jim. Spent all my holidays at Kyogle . . . horses and hot days in the bush. Once, when I was a kid half the town was burnt down. Well, we'll all be seeing snow this winter, I expect.'

The following morning Japanese guards lined them up and issued them with one blanket apiece. The naked used their blankets as cloaks or wore them toga fashion. Then they were marched to a train and ordered aboard by an officer who repeated something they took to be 'Tokyo'.

Rowley's impressions of the train journey—entering a big city that seemed to continue mile after mile, the final stopping in shunting yards—were jumbled and vague. The shunting yards seemed to be associated with a dock area, although there were so many acres of lines and buildings that no actual docks or harbour could be seen. From the talk among their guards he gathered that they had stopped in Yokohama. His knowledge of Japanese geography was sketchy but he thought that Tokyo and Yokohama must merge. It was the main industrial port of Japan, which would account for the impression he had of vast dockyards.

The prisoners were unloaded and marched through the streets to a depot; a shuffling, limping, shivering mob, clutching blankets around their bodies; the fittest supporting or carrying the injured, providing a spectacle for the local population about the docks. They were greeted with laughs and jeers. Despite the submarine blockade the Japs still seemed confident they were the top dogs in this war.

At the Yokohama depot the survivors were given an issue of clothing; green jacket and slacks and shirt. They were split into three groups. Three hundred British were sent into Tokyo. Fifty of the Australians—including Duncan, Mills, Major, Johnson, Thomas, McGovern and Campbell of the *Perth*—were drafted into a Yokohama prison camp. When he saw that the Australian group was being split up Rowley requested an interview with the guard commander and protested strongly. He was deeply convinced that, after so many tribulations, the retention of their identity as a group was their only security in the land of the enemy and was vital to their survival.

But the Japanese commander waved aside his protest as if it were some trivial irritant, the buzzing of a fly. Dock workers, he stated, were needed in Yokohama. The men would be well treated. There was no need for concern. He was busy. Captain Richards must be ready with the rest of his men to entrain.

On 3 October they reached the next and final stop, which was the harbour town of Sakata on the north-west coast. The third group, the remaining 261 British and twenty-nine Australians, were off-loaded near the docks. Their temporary quarters were in a little school. Rowley tried to shake off lassitude and take stock of their group. The seriously ill British officer had been left in Moji, along with eight British ORs and one Australian, Sapper Burridge. All were too sick to travel. Captain Wilkie had remained in Yokohama, but the other British officers were with the Sakata party. Rowley and the other twenty-eight Australians were a pitifully small remnant of the original Japan Party.

The school was unheated and the only bedding was matting on the floor. Men sat around with blankets wrapped about their bodies and stared dull-eyed. A ration of sloppy stew, mainly barley with a few vegetables but no meat, was brought to them by guards who were civilians. With a little warm food in his stomach Rowley felt the need to take some sort of action. What action was not important as long as he did something to stir the men out of their apathy. He called one of the guards, stated that he was a medical officer and needed a room in which he could treat the sick. The civilian grunted and went away. Rowley wondered if he had understood or would bother to transmit the request to someone with authority.

Within a short time a little Jap civilian who limped on a withered leg typical of the deformity inflicted by poliomyelitis appeared with the guard who directed him towards Rowley. He introduced himself as Matsumoto-san. He was the civilian medical orderly for the Sakata prisoner-of-war camp which was to be administered by the Nippon Tsun company. He seemed friendly and sympathetic and immediately agreed that the prisoner doctors should be given a room where they could commence treatment.

'Captain Roulston,' Rowley called out, 'we can inspect our medical quarters.'

As he and Roulston followed Matsumoto out of the school assembly hall that was to be part of their sleeping quarters, Rowley noted that the British officers were on their feet and moving among their men, intent on some form of organisation. Warrant Officer Keith Martin, the 2 I/C of 35 Kumi and Kitch Loughnan had gathered the Australians around and were talking to them. They looked more alert now.

The room to which Matsumoto took them was sufficiently spacious to serve as a hospital which would comfortably accommodate ten patients, leaving one end to serve as a medical inspection and treatment room.

Rowley nodded and smiled at Matsumoto. '*Joto des. Arigato.*' (It is good. Thank you.) He wondered what he should do next, but his thinking was sluggish. For the first time since becoming a prisoner of war he admitted to himself that he felt almost too exhausted to make an effort to reorganise. The disasters had come one on top of the other like dumper waves in the surf, pounding, grinding, suffocating before body or mind had had time to recover from the previous one. Yet he knew that survival for all of them lay in reorganisation: the effort of discipline of many shattered selves to become a cohesive group again, each part lending strength to and drawing strength from the whole.

Roulston, he realised, was also making an effort to start functioning again.

He was asking Matsumoto questions about the issue of medical supplies. Matsumoto said some supplies would be made available on the next day.

'Nothing more we can do now, old chap,' Roulston said with an effort at heartiness. 'But we should be able to hold a sick parade tomorrow. Get back into the swim of things.'

Rowley blinked. He saw himself holding sick parade in the jungle, then on the *Rakuyo Maru* with an orderly at one side of him calling from the nominal roll and an orderly on the other side noting treatment.

The roll! Of course.

'We can hold a sick parade now,' he said. 'We must make up a new nominal roll. We have no records.'

A bark of laughter exploded from Roulston, the first real break in the gloom that had enveloped him since Rowley found him among the sufferers on the deck of the tanker. 'Rowley Richards, compulsive record keeper. Right you are, old chap. We need a nominal roll.'

'We need a roll book and writing materials,' Rowley told him.

Matsumoto supplied pencils, regretted that he was unable to supply a proper roll book but offered a small black-covered notebook that was his personal property. Captain Peirce organised the British and Warrant Officer Keith Martin the Australians into the order of their POW numbers and lined them up outside the medical inspection room. Rowley, sitting cross-legged on the matting with the notebook on his lap, slapped his knee gleefully. 'Right Roulston. We are in business again. Start bringing them in.'

Working together, Rowley and Roulston not only compiled the roll, they recorded the height, chest measurement, weight and foot measurement, and current illness of each man.

'What's measuring me foot got to do with treating me tinea, sir,' Jackie McKone wanted to know.

'So we can get you some boots, of course,' Rowley told him. 'Haven't I drummed into you for years that you must not go barefooted in the East.'

'Yes, sir. Sorry, sir. Somehow I lost me boots in a shipwreck. Took 'em off to swim.'

'Careless bastard, aren't you? Could have tied them round your neck. At least keep your socks on inside, until we get you boots.'

Later, when the men had returned to their sleeping quarters much more cheerful and alert, Rowley sat alone in the medical inspection room and considered the names of the Australians in the nominal roll: CSM Keith Martin, age 36; Sgt Ernie Wade, 32; Sgt Ross Dunbar, 25; Sgt Arthur (Butch) Hall, 31; Sgt Geoff Macaulay, 37; Jimmy Mullens, 23; Russell Savage, 23; Kitchener Loughnan, 30; Clarrie Wilson, 27; Lindsay Savage, 35; David Clark, 23; Stan Manning, 27; Walter Dorman, 39; Jackie McKone, 41; Ian MacDiarmid, 24; Tommy Moxham, 29; Harry Hodgson, 28; Steve Weekes, 29; Herb Wall, 22; Jasper Schwerdt, 37; Frank Attrill, 25; Eddie Anderson, 28; Jack Smith, 36; 'Titch' Lemin, 24; Teddie Kelk, 42; Alick Robinson, 30; Chris Dowd, 37; Jack Wade, 29. He knew everyone on that list well, their strengths and their weaknesses, and every one of them was important to him. 'As no man is an island, complete unto himself . . .' Every one of these was a part of the Main. Would the bell toll for some of them before the war was over?

Their average loss in weight since leaving Singapore was ten kilos, and that was ten kilos below their average weight on the line and aggravated by loss of beriberi oedema fluid. The main medical problems were boils and skin infections (55%), oedema due to beriberi and protein deficiency (35%), malaria (14%). Most were suffering from more than one condition and they were about to face a winter in Japan.

On the following day the prisoners were paraded before Lieutenant Watanabe, a smart, rather good type of Imperial Japanese Army officer. He explained that they were to be moved shortly into a rice warehouse, part of the premises of a nearby sake factory that was to be their permanent accommodation. They would receive a further issue of clothing and eating utensils. The rations, as outlined by Watanabe, seemed reasonable. In response to a question from Captain Peirce the Japanese officer even promised an issue of boots. The work quotas and hours of work were stated and seemed fair enough. If conditions were as promised Rowley thought, they would all have a good chance of surviving.

Nippon Tsun, a large stevedoring and manufacturing conglomerate, was the successful tenderer that had contracted with the Imperial Japanese Army to provide this Sakata group of its war prisoners with food, clothing and shelter in exchange for labour. The labour required for the several divisions of Nippon Tsun was: *Tekosha*, a chemical factory—filling and moving trucks in and around the factory; *Shotan*—unloading coal from barges and loading into railway wagons; *Ikada*—dismantling rafts of logs in the water and loading timber onto railway wagons. The working day was from 0630 hours to 1700 hours, ten and a half hours with a break of one and half hours for the midday meal and half-hour smoko breaks in the morning and afternoon.

Only eight actual working hours! It would have been a reasonable working day for men in good health, almost a labourer's paradise for men who had experienced the Burma-Thailand railway. But none of the survivors was in good health. Worn out by long ill-treatment, they needed food and rest. The diet was insufficient, about a pint of a sloppy mixture of rice and vegetables at each meal, the vegetables consisting mostly of some unidentifiable dried chunks and some seaweed. There was no meat.

'That has to be one of the most obnoxious things I've tasted,' Rowley told Roulston, holding aloft in his wooden spoon a pale-coloured chunk that he had fished out of his bowl of rice. 'I much prefer the seaweed.'

Roulston nodded sagely. 'It has a piquant flavour,' he observed. 'And it is said to be nutritious. I say, Richards, these Nippon Tsun fellows are falling down on their bargain as far as rations go. We are not getting the ration quoted to us by Lieutenant Watanabe. What do you think?'

'I think we should complain,' Rowley told him. 'But politely. These people are a whole lot more formal than the Jeep and the Storm Trooper. Let's start with the Nippon Tsun bloke.'

Through the interpreter, Rowley and Roulston arranged an interview with the camp administrator. The interpreter was a soldier from a local army unit but the camp administrator was a civilian, a Nippon Tsun employee. His neat suit and spotless shirt contrasted sharply with the shabby gear of the Australian and British officers who waited on him. They wore the same winter issue clothing as all prisoners in the camp. Their suits were a cheap worsted that did little to keep out the cold, their shirts and boots were second-hand, as were their underpants. The cotton socks had been new when issued but one pair of socks quickly rubbed into holes and they had no darning materials, so their feet became blistered, then calloused, in the ill-fitting boots. At least the captured British overcoats given them were reasonably thick and warm if not the height of fashion.

They stated their complaints then waited while the interpreter translated. The civilian looked at them curiously, then shrugged and gave a short, coughing laugh. He spoke to the interpreter.

'The administrator appreciates your concern that the men are unable to fulfil their work quota because of ill-health and will give sympathetic consideration to your request for further nourishing food so that the men may be restored to full health,' the interpreter said.

As the administrator then became immersed in reading some papers, Rowley and Roulston bowed themselves out.

'Is that really what he said?' Roulston asked.

'As far as I could follow that was more or less it. But he started by saying "More food? There isn't any." And he's a bloody liar. One of my blokes saw Nippon Tsun people loading rice bags and some boxes—dried fish, he thought—onto a truck just after the last delivery of supplies. It went out of the compound.'

'I've heard the same thing,' Roulston noted. 'They are helping themselves to our supplies. And the blighter laughed. Probably thinks it's a big joke when we ask for more.'

The rations did not improve and Rowley's hospital section admitted three more cases of gastric complaints due to the imbalance of the diet, which almost completely lacked protein. He had two dysentery patients who were deteriorating alarmingly. One was a Britisher, the other young Jimmy Mullens. Both were passing blood but he couldn't get special invalid food for them because the prisoners did not have a kitchen. The cooking was done somewhere outside the camp and delivered to the compound. There was little he could do except give them boiled water, soup that was strained from the sloppy rice-stew mix, and keep them resting in bed and as warm as possible in a drafty, unheated hut. Nippon Tsun had been asked to supply stoves for the billets and the hospital but these had not been produced.

'I'd say Nippon Tsun are out to save their yen at the expense of POWs. If they aren't fulfilling their contract with the IJA it's time we started letting the military administration know—in writing.'

Rowley borrowed desk space and writing equipment from Captain Barrett, who had been designated works officer and handled all clerical duties. With suggestions from Roulston and Pote-Hunt, who was adjutant and quartermaster, he began drafting a letter to the military commander of the Sakata camp, Second-Lieutenant Watanabe.

'The Officers of the Sakata POW camp respectfully wish'—the polite, formal approach was the right touch; officers in Japan were sticklers for etiquette—'to make the following report on the health of the men under their care and the difficulty of maintaining working figures in spite of sending the light sick to work.'

Rowley visualised the state of the men in the Sakata camp. Physically, all 290 of them suffered extremely from exhaustion aggravated by starvation, malaria, dysentery and avitaminosis. Their weight loss ranged from ten to as much as twenty kilograms below their already reduced POW railway weight. They reminded him of the scrawny Burmese cattle. Not one of them would make a decent meal for a hungry cannibal. They needed rest and nourishing food. He wrote on:

1. The condition of the men on arrival in Japan following their shipwreck and exposure was at a very low level.
2. For the men to regain their normal working health it was necessary for them to be sent to a Rest Camp and built up with plentiful, wholesome food.
3. On arrival at Sakata the men were rested for three days and then sent to light work. The food was wholesome . . . [Rowley supposed it was by Japanese standards, if the body was conditioned to seaweed and the ghastly dried vegetable stuff] but the quantity insufficient for the men to do any work and regain their strength at the same time.
4. Since then the work has increased but the food ration has remained unaltered. For example, at Shotan the daily task performed has been increased from five to fifteen trucks. [Did Watanabe, Rowley wondered, know that Nippon Tsun were increasing the workload? Well, that paragraph would do to inform him and suggest, in politely oblique Japanese fashion, that more work rated more food.]
5. These men, even when doing heavy work in Burma and Thailand, were accustomed to receive 800 g of rice daily which allowed them to have about 750 cc of cooked rice as well as 500 cc of meat and vegetable stew at each meal, 500 cc of tea was also issued at each meal. Our present ration allows for approximately 500 to 700 cc of 'stew' per meal. This ration is far short of the scale which was mentioned to us by the contractor and which would be quite adequate . . .

'That's telling him the mean rotters are short-rationing us,' Jim Roulston approved, looking over Rowley's shoulder.

> . . . Until such time as this ration scale becomes effective we would request an issue of eight small tubs of dry rice in addition to the sixteen small tubs of 'stew' now being issued. Further issues of fruit would be greatly appreciated.

Captain Peirce came into the tiny administration office, his coat collar turned up against the cold. The mid-October days were dry and clear—at least there was not the incessant, mind-numbing rain to contend with—but it was cold, with promise of the coming of the 'purifying' snows and there was a cruel bite to the wind.

'A hot cup of tea would be wonderful,' Peirce said. 'Why the hell did they issue us with kettles but no mugs? I've just been having a word with the Nippon Tsun fellow about the delay in building kitchens for us. He shrugs and grins and seems to think I'm joking.'

'We've had a word to him about rations. He took that as a joke too,' Roulston told his fellow officer. 'Richards is composing a letter to Watanabe.'

'Well, ask him if we can borrow their kuallies to make tea.'

Rowley continued with his draft . . .

> We would also request that the kuallies in the present camp be made available to us for the preparation of hot drinks at meal times.
>
> 6. Further reasons for the high sick rate and the reduction in the working figures are that the skin complaints contracted after the shipwreck have not yet cleared up and that the malaria recurrence rate is increasing. The former is due to a shortage of gauze bandages and hot water and the latter due to insufficient quinine therapy. It is thought that necessary medical supplies may be obtained from POW hospitals in Japan which have received International Red Cross medical supplies. [That should jog him into remembering that Red Cross supplies are supposed to go to us, not be confiscated by the IJA.]
>
> The officers wish to reassure you that the men can and will work to your satisfaction if given adequate food and medical supplies.

Rowley handed the draft to Peirce. 'That should do for a starter, if you approve.'

Peirce, as the Royal Artillery regular officer in charge of the prisoners of war, and Rowley as the Australian Medical Corps officer, signed the letter. Peirce took it for delivery to Watanabe and they awaited his reactions.

The camp at Sakata was a barbed-wire enclosed compound, about seventy-five by fifty yards, containing the rice storehouse which had been converted into billets by the construction of a platform eight feet from the floor and reached by vertical ladders. The wooden floors had been covered with the Japanese straw mats, used for sleeping, and an issue of three blankets per man had been made when clothing was issued. The sleeping space was adequate, six feet by three for each man with eight feet of headroom, but the billets were draughty and no form of heating had been supplied although stoves had been promised.

Two small rooms had been allotted as Medical Inspection Room (the British term for RAP) and hospital. This was the joint domain of Jim Roulston and Rowley, who together cared for the entire group of prisoners. They had three British POW medical orderlies for whom they had arranged sleeping accommodation in the MIR so that there would be constant attendance on the sick.

The sliding doors on the exposed north and west sides of the building and between the MIR and hospital were ill-fitting and the cold wind came through the gaps. The sick had only the several layers of matting on the wooden floors and their issue of three thin blankets for warmth. Although the hospital could comfortably accommodate only ten patients there were still sixteen hospitalised in mid-October, with malaria, dysentery, beriberi and, in two cases, corneal ulceration—caused by immersion in oily salt water—so bad that Rowley and Roulston had little hope of saving the sight in one eye of each of these British prisoners.

The malaria cases were responding to a limited course of quinine, which had been included in a case of medical supplies issued by the IJA on the group's arrival at Sakata. The case included a stethoscope, scalpels, scissors, probes, forceps and some medicines; aspirin, quinine, sulphanilamide, Mercurochrome (desperately needed for the treatment of septic sores), sulphur, soda bicarb and icthyol.

It was a far better dispensary than anything Rowley had had on the railway but was still too limited.

There were no amoebicides, which he needed badly for the treatment of amoebic dysentery. He was fighting now for the lives of Jimmy Mullens and the Britisher, who no longer had the strength to battle their disease without adequate drugs. Rowley asked Matsumoto and the two civilian doctors, who visited the camp once a week, for the amoebicides. The doctors seemed genuinely helpful, but they shook their heads regretfully and explained that civilian medical supplies were limited, the army having priority on all supplies. There were no amoebicides available.

Jimmy Mullens died quietly after a severe haemorrhage on 20 October 1944. He had been fated never to see snow falling. Pulling a blanket up over the face of the young surfer Rowley remembered vividly how he had swum about so cheerfully after the shipwreck, getting other men into the lifeboat; how he had stood by the jury-rigged sail in the boat and declared he was game to sail to China. Now that vigorous young body that had been dedicated to water would be consigned to fire. There would be nothing left but a small box of ashes which would remain in Japanese custody until the war was over. Rowley hoped he would survive to see those ashes returned to Manly.

Why had Jimmy succumbed to the dysentery when he had a young healthy body and a great spirit? Rowley frowned. Jimmy had been prodigal with his physical strength. He had never learned to harbour his resources, to protect himself. Recalling the deaths on the line Rowley realised that the older men, whose reserves were spent, and the very young, who had not learned to conserve their resources, were the most frequent victims. The twenty-five to early-thirty age group seemed generally the best equipped to survive. Self-protection as well as self-discipline was part of the survival equation.

The Britisher died the day after Jimmy Mullens.

All the prisoners were becoming weaker. The diet was not enough to build up their strength. And it was already so cold—a terrible cold that sapped their minimal reserves of energy. How were they to survive the winter?

12

The Survivors

The first snow fell on 24 November. For days it fell steadily and the compound was thickly covered, but men had to walk from the billets to the latrines with bare feet. The second-hand boots had rotted and split quickly and Nippon Tsun had replaced them with straw-rope sandals which were generally too small. These had fallen to pieces in a matter of days and then the civilian administrator had produced a supply of straw, saying that the prisoners could make their own footware.

As far as Rowley was concerned, winter in Japan—and it was the worst in that part of the country for seventy years—was merely exchanging cholera, flies, heat and torrential rains for pneumonia, aching cold and snow. Beriberi and cycles of malaria were still with them, and so was the amoebic dysentery despite the 'purifying' snows. Instead of Korean guards forcing them to labour on a railway they had the Nippon Tsun company's employees as overseers.

Before Christmas there were seven cases of pneumonia in the hospital. Sulphapyridine, essential in the treatment of pneumonia, was absolutely unobtainable although Lieutenant Watanabe said he had done his utmost. Rowley believed him. Watanabe had succeeded in getting a supply of dysentery powders, although emetine for the amoebic dysentery was, he regretted, unavailable; also extra quinine. He had exerted pressure in the right places for the Nippon Tsun contractors to build kitchens for the prisoners. The preparation of food was improved, although the kitchens had no window panes and the cooks had to work in their overcoats, and the kettles for tea-making were steaming away at last.

Captain Peirce asked the civilian administrator for stoves to heat the billets and hospital and again had the impression that the request was treated as a joke. He asked Lieutenant Watanabe. Extra blankets, a charcoal brazier and a supply of charcoal were sent to the hospital. Rowley and Roulston, with their orderlies, hung the blankets around the walls to keep out the wind. With the brazier glowing, the hospital was relatively warm and they had hopes of saving some of the pneumonia patients. Despite their efforts one of the British died before Christmas.

Beriberi was on the increase and there were deaths among the British from cardiac beriberi, death taking place within an hour of collapse. Rowley noted that marked dehydration was a feature of beriberi once oedema had been reduced by rest, limited intake of water and salt, and improved diet. Vitamin B preparations were supplied for the hospital, but in the absence of protein they seemed to have little effect.

'Remember Padre Pete?' Kitch Loughnan said to Rowley on Christmas morning 1944. They were huddled in their overcoats at the medical inspection room end of the hospital, grateful for the warmth of the brazier.

Rowley smiled, remembering the snake in the grass near the cholera isolation hut. 'How could I forget him?'

'I mean . . . when the boys talked about being home for Anzac Day, or home for Christmas, the most important day in the Christian calender, Padre Pete used to say, "Oh, I don't think so. Not this Christmas. The war could go on for years, you know. It could go on for five years."

'No wonder we called him Pessimistic Pete. Maybe he was right. Anyway, we won't be home this Christmas or New Year, or even Anzac day, I think.'

Kitch's long face was grave, its characteristic grin missing. There was a crease between his eyebrows and his cheeks were hollowed with anxiety.

'Come on, Kitch, you aren't Pessimistic Pete. We'll be home for Anzac day.' Rowley encouraged.

'Don't have me on, mate. I'm a big boy, now. We're all weak. How long can we last?'

'Seriously? I think we can outlast the Japs. The sub blockade is hurting them. Look at Sakata. Working here we can see what it's like. Food, medical supplies, clothes, fuel, everything is short for the civilians. The Army grabs everything. Jap troops could dodge around the Pacific, fighting for years, but when the homeland hurts it's a different matter. Japan hasn't natural resources, Kitch, and now it's a country under seige. We just have to hang on.'

'That could be right.' Kitch nodded thoughtfully, studying his big hands. 'The only meat that goes through the abattoirs is the occasional horse that breaks a leg on the ice, which is meat for the entire town. They have whale meat—fishy steak!—but the war at sea is affecting that supply too. Funny, there I am working in the abattoirs and I never thought of that angle before. How bad it is for them!'

As experienced cattle men, Kitch, Lindsay Savage and Butch Hall had been put to work in the local abattoirs. The slaughterman there was Takahashi, who was a friend of Matsumoto. He shared tea with them and did not press them to work too hard. Just that little bit of protection improved their chance of survival.

Kitch looked happier. 'Of course we can hang on. She'll be right. You can't kill a good Aussie with a meat-axe. Rowley, are you going to give us a song at the concert?'

Rowley's grin, rarer in these days of biting cold and hunger, flickered. 'That would be sure to kill you. I wouldn't prescribe my singing as a tonic for anyone.'

The concert was organised by the Japanese civilian staff. They had borrowed a piano and other instruments and arranged a stage inside the billets. The families of some of the civilians attended and even performed. It was cold in the billet and everyone wore overcoats but the prisoners applauded enthusiastically, amazed that the civilian staff had wanted to give them a good time.

They had planned a small gesture of gratitude. There had been an issue of Red Cross food parcels, one per seven men, and the prisoners, instead of eating all their few squares of chocolate, had kept some as surprise gifts for the children of the Japanese who had arranged the concert. As Rowley watched the faces of the black-haired, slant-eyed children who were not his enemies and saw the near reverence with which they accepted the little slabs of chocolate he felt glad that his men had had a gift to offer.

He discovered that Matsumoto was beside him. The little civilian medical orderly who supplied the camp with scraps of war news and was consistently helpful in attempting to obtain more medical supplies had brought a gift of extra charcoal for the hospital brazier. He was shifting his feet uneasily and Rowley thought that his withered leg must be paining him.

Matsumoto coughed diffidently. 'Doctor-san, I regret my child could not attend with me,' he said. 'He has been ill with a chill. We have not had such a cold winter in my lifetime. If I might ask a favour . . . could I take a square of chocolate for my child? He is five years old and has never tasted chocolate.'

Rowley looked at his empty hands. Then he remembered that there was a Red Cross parcel in the hospital, issue for the patients who were too sick to attend the concert.

'Wait here, Matsumoto-san.'

In minutes he was back with a small block. 'Take that to your child with the greetings of the Australian soldiers in the hospital,' he said formally.

He thought that there had to be better ways of solving world problems than going to war and turning men into killing machines, teaching them to hate. You realised in the end that no one really won, and all civilians suffered.

But you had to go through a war to know that. Succeeding generations didn't know it. He hadn't known himself on 3 September 1939 when he was a young militia officer listening to the official declaration of war. Now he knew why the veterans of the Great War had been so sad.

Men were aggressive animals and somewhere there were always aggressive, power-hungry brutes dreaming of conquest. Decent people and decent attitudes were pushed into the background, often because the decent ones were afraid of asserting their decency in the face of the greater power of the brutes.

A people attacked had to defend themselves and their way of life. But war was dehumanising, even if you believed you were in the right. Brute instincts were unleashed. In the terrible moral chaos things came to the surface like the murder of Sergeant O'Donnell, the bashing of Jim Lynch, the torture of Colonel Williams. Yet, with all the primeval muck of the spirit that was stirred up, there were still strong, clear currents: the guard who had helped save Bill Drower's clothes from burning; the Australian digger escorting a Japanese prostitute aboard a lifeboat; the amazing instance of Japanese naval personnel returning their ships into submarine-infested waters to pick up castaway prisoners of war.

These were things to remember, not only at Christmas but whenever the going was tough. They made life seem worth living. And it was only if you thought life was worth living that you maintained the will to survive. You could go on until the flesh gave up.

Rowley's eyes went to his group of Australians, all bunched together, supporting each other in their fight to retain pride, courage and humanity. Although he shared responsibility with Roulston for all the medical work in the camp, that minority group was his special concern. They had come a long way together, and he was proud to be one of them.

Jackie McKone was not with them for the Christmas party, neither was young Russell Savage, Lindsay's cousin and one of the babies of the group since Jimmy Mullens had died. Both were very ill and had been sent to the Japanese military hospital in Tokyo. It was a protection that McKone, now aged forty-one, needed badly. Rowley hoped that the Japanese would keep him in Tokyo. Back in the camp he wouldn't stand much chance.

A few days after Christmas Kitch and Lindsay returned from the abattoirs with a couple of ribs of horse meat, a gift from Takahashi. The slaughterman had sliced away the lowest ribs when quartering a carcase. It was a clever trick. He was still able to account for all four quarters and the ribs were not missed; if they had been he could well have been lynched. Thanks to Takahashi there was nourishing broth for the hospital patients and a gravy with shreds of meat to flavour the rice.

Rowley felt that he should try to keep up some form of mental stimulation. There were no books available for the prisoners so he began to compile a list of philosophers remembered from his readings and discussions with John Shaw while he was on the railway. He arranged them chronologically: Socrates, Plato, Aristotle, Epicurus . . . Roger Bacon, Francis Bacon . . . Spinoza, Kant, Hegel, Schopenhauer and others. Although he had never had a good memory for quotations, he tried to remember and quote and was surprised by what he had retained.

He wrote in the notebook provided by Matsumoto:

Seven Sages of Greece—

Solon	—Know thyself.
Chilo	—Consider the end.
Thales	—Whoso hateth suretyship is sure.
Bias	—Most men are bad.
Cleobulus	—Avoid extremes.
Pittacus	—Seize Time by the forelock.
Periander	—Nothing is impossible to industry.

Thoreau	—The most glorious fact in my experience is not anything I have done or may hope to do, but a transient thought or vision or dream which I have had.
Berkely	—The outer, as regards our knowledge of it, is within.
Goethe	—Think of living! Thy life is no idle dream but a solemn

	reality. It is thy own, it is all thou hast to confront eternity with.
Georgias	—Nothing exists, or if something exists it cannot be known, or if it is known it cannot be communicated.
Emerson	—The highest price we can pay for a thing is to ask for it.
Schopenhauer	—The world is neither with you nor against you, but is putty in your hands and may be heaven or hell according to your *will.*
Jerome K. Jerome	—Love = Respect, Admiration, Affection.

After sick parade he sat by the hospital brazier and continued to write:

> The Magnanimous Man—Plato. The magnanimous man will behave with moderation under both good fortune and bad. He will know how to be exalted and how to be abased. He will be neither delighted with success nor grieved by failure. He will neither shun danger nor seek it . . .

It's good, he thought. I must read that to Kitch when he comes back from the abattoirs.

Pain gripped his belly and he put down the pen. He would have to go out into the cold again and walk the snow-covered path to the latrines. The diarrhoea was getting worse and if it didn't ease up Roulston would have to take over all the medical work for a few days. Although he hated to admit it, he was sick.

His legs felt like rubber and he collapsed before reaching the latrines. Someone, one of the cooks he thought, helped him up. 'Want a hand to reach the latrine, Doc?'

When Kitch returned from work he dropped into the hospital and found Rowley, supported by Roulston, hobbling out on another trip to the latrines. It was a common sight to see men supporting each other along that snow-covered path but Kitch had a better solution.

'Rowley's sick? Here . . . give him to me.'

He picked Rowley up like a babe in arms and carried him.

Kitch asked to be assigned to medical orderly duties for the duration of Rowley's illness, carried him to and from the latrines and fed him hot soup sent in from the kitchen. The soup was good, a meat stock with actual shreds of flesh in it. It was being served to all the hospital patients. When he was strong enough to wonder about it Rowley decided that Takahashi must be pinching more horse ribs for the camp. He hoped Takahashi would not take too many chances. The meat-hungry citizens of Sakata would not be kind if they found out what their slaughterman was doing.

By late February Rowley was well enough to resume duties and walk the path to the latrines by himself. Kitch resumed work in the abattoirs. The thaw was beginning and the high banks of snow were melting slowly. Rowley was walking along the path from hospital to kitchens when he saw that an object was protruding from the snow. It was a small, pointed, furry face. He looked more closely and thought he was hallucinating, but he was not. It was undoubtedly the face of a dog. Some four inches beneath the face two little furred paws were visible. The poor thing was frozen. He wondered how it had come to be trapped in the snow bank. Then he saw that there was another dog muzzle protruding a couple of feet from the first one.

Rowley dug into the snow. The fur ended at the neck and paws. As far as he could see without digging too deeply, the body of the dog had been skinned like a rabbit. A horrible suspicion formed. He felt like dashing to the kitchens to demand confirmation from the cooks, or back into the hospital to tax Roulston. Then he decided he was being naive. He would wait until Kitch, Lindsay and Butch returned from the abattoirs and get an explanation from them.

He waited until he got them alone. Kitch and Lindsay both looked upset. It was left to Butch to explain.

'The Jap civil administration reckoned dogs were eating valuable food and issued an edict that they were to be rounded up and slaughtered. Their skins are being tanned but the edict didn't say what was to be done with the carcases. The Japs won't eat them—they aren't hungry enough yet—so Takahashi said we could take them. We've been bringing them back and deep-freezing them in the snow banks. The cooks dig them out as they need them. We won't be able to keep them now the thaw has started.'

'Those poor little faces looking out of the snow!' Kitch burst out. 'Dogs have always been my mates. Hell, Rowley, how are we ever going to tell people . . . make them understand?'

'Everyone in the camp has been eating dog,' Lindsay said defensively. 'The cooks do a really good job. You've been eating the meat, Rowley. It's all right, isn't it?'

Rowley considered. 'All right? I've thought the extra meat ration was delicious. Do you remember that incident at the River Valley Road camp when some Dutch knocked off someone's pet dog and ate it? Our blokes were ready to riot about that business. I'd been hungry on the railway yet I was horrified at the thought of anyone eating dog. Well, the meat's all right. I'll go on eating it.'

Lindsay sighed with relief. 'That's good-o! I still can't help feeling bad about it. But we need the food.'

He pulled up his thin trousers and looked with distaste at his swollen legs. When he poked the unhealthy flesh the dent made by his finger remained. 'Will these legs ever be normal again? Will I be able to get rid of the oedema?'

'After the war—with better rations—we'll all be right,' Rowley said.

On 1 March Lindsay Savage collapsed and died, a victim of malaria, beriberi and pneumonia. He was thirty-five years old. Another box of ashes was stored for return to the homeland when the war was over. Kitch said little except, 'I'll miss those arguments about cattle,' but he plodded grimly to the abattoirs each day and did not smile very often.

Sergeant Ross Dunbar, aged twenty-five and a chronic sufferer from amoebic dysentery, beriberi and malaria, was sent to the hospital in Tokyo. Russell Savage had returned from the hospital bringing news of the Australians who had been sent to Yokohama, but Jackie McKone was still hospitalised. 'He's battling on but still weak,' Russell told Rowley. 'He's sensible enough to keep on being weak, too. Jack will stay in hospital for the duration if he can.'

Before the snow had completely thawed the supply of dog carcases had run out. Men working on the docks gathered seaweed from the wooden piles and brought it back to the kitchen to be added to the meagre stew of barley, rice and dried vegetable.

The snows melted and the spring came, but the Japanese did not seem to be much cheered by the beauties of spring in their cherry-blossom land. Food and all domestic commodities were in short supply. The war news featured the losses inflicted on the enemy by Japanese troops . . . while making strategic withdrawals.

'I think it is more grave than we are told,' Matsumoto confided to Rowley.

On 15 April 1945 the administration of the camp was changed, coming under the direct control of the Imperial Japanese Army. Rowley hoped they had finally caught up with the yen-pinching tricks of Nippon Tsun, whose employees were replaced by ex-servicemen, with an IJA officer in command. The new personnel were a sergeant in charge of supplies, a sergeant in charge of medical arrangements, one IJA medical orderly and a corporal in charge of administration.

'I must go to other employment for my organisation, now.' Matsumoto said. 'I am sorry to say goodbye to you, Doctor-san.'

'I am sorry too,' Rowley replied, and was astonished to find that he meant it.

'My friend Takahashi will give you news of me through your friend, Kitch. If I can be of assistance in any way he will tell me.'

'Thank you.'

They both bowed formally and Matsumoto limped out of the hospital.

The new camp commander was Lieutenant Mori, who was sympathetic as Watanabe had been, and unhampered by having to deal with Nippon Tsun he got things done. Stoves were supplied for the billets. They were not provided with chimneys and consequently smoked badly, but a warm smoky billet was preferable to a cold one. The spring was chilly and there were some deaths among the British from pneumonia. Roulston was unflagging in his care for the sick, seemingly tireless but with the automatic tirelessness of a robot. He withdrew more and more into the defense of his English reserve. Three issues of Red Cross parcels were made during the summer—one per seven men on each occasion.

The ladders reaching to the second-floor sleeping area in the billets were converted to inclined stairways. The bathhouse was completed and supplied with warm water and the men were able to have their first warm bath since leaving Singapore.

During the worst of the winter months, with only one cold tap for 300 men, Rowley had realised that it would be almost impossible to carry on preventive hygiene methods. Men couldn't be expected to shave with icy water or strip to the pelt for skin parades. He was gratefully surprised when he found that his remaining Australians, despite physical weakness, were organising skin inspections among themselves, and from the time the kettles had been made available, had insisted on getting boiling water from the kitchens to sterilise their eating tools. They shaved, too, although beards would have been some protection against the icy wind.

'I'd sooner have a cold face than a face bitten all over by nits,' Butch Hall said. 'Lice in your clothes are bad enough.'

They huddled together on their sleeping platforms at night and held nit-picking sessions—delousing their clothes. Usually someone would start up a quiz like the ones that had been popular on the line, but the answers came more slowly and memories were noticeably less reliable. It seemed that their minds were freezing too, and mental exertion, like physical, was an effort. But they persevered.

Most of them were now free of bowel complaints and the incidence of serious skin sores was small although they worked bare foot in slush and water at the coal skips and at the wharves. The malaria sufferers (and most of them suffered cyclic malaria), were responding well to courses of quinine.

Jackie McKone was returned from Tokyo hospital, fairly fit, on 20 May. He had met some people from the Yokohama camp in hospital and the news was that the *Perth* men who had been in their lifeboat were fit. Rowley asked to have McKone assigned to the hospital as a medical orderly, telling the sergeant in charge of medical arrangements that Jackie was his most experienced orderly and was needed at the hospital. In fact he had to be continually protected if he was going to make it. At forty-one he was the second oldest of their group. Kelk, who was forty-two, was the oldest.

On 30 May two of the old men, Lance-Corporal Dorman, thirty-nine and Sapper Dowd, thirty-seven died of pneumonia. Eleven days later Private Schwerdt, who was also thirty-seven, died. There were now five boxes of ashes to be returned to Australia. Rowley remembered Roulston, on the whaling mother ship, talking about ten little nigger boys. They were now twenty-four . . . Oh, damn Roulston!

Although June was officially summer and the billets were warmer despite the rains which had commenced, Kitch had been getting thinner and grimmer since Lindsay had died, and developed a chest cold. He said nothing about the feeling of constriction in his chest and the slight fever. He went to the abattoirs as usual but when he returned to the camp, flushed and shivering, Rowley ordered him into the hospital.

'It's just a bit of a cold,' he protested, hugging his overcoat around him although the evening was fairly mild. 'I have some cold sores on the mouth.'

Rowley checked the herpes and moved his Japanese stethoscope over Kitch's broad back. The dry, rustling rales were clearly audible. Virulent pneumococci were rioting in Kitch's left lung. By the time Rowley and McKone had settled

Kitch on a sleeping mat near the lighted brazier, bundled in blankets, his temperature had risen. His face was hot, flushed and dry. During the night his breathing became rapid and shallow and he grunted occasionally with the pain of respiration. Rowley took turns with McKone in snatching a little sleep but even in his sleep was aware of the laboured breathing.

Roulston offered to take over during the next day but Rowley shook his head. 'I'll look after Kitch.'

'You'll collapse, old man.'

'By God, I won't.'

He bathed Kitch's hot face and coaxed him to take drinks of warm water. The temperature continued to climb and the sputum Kitch coughed up was rust coloured. With antibacterial drugs the pneumococci could have been knocked out within forty-eight hours, but in the absence of sulphapyridine the disease had to take its old-fashioned course. The crisis usually occurred about the seventh to tenth day. Then the patient began to sweat and the temperature fell rapidly: that is, if the patient was going to recover.

On the third day Roulston again suggested that he should take over.

'You can't go on like this, Richards. And there is nothing that you can do that I can't.'

'I can stay with him,' Rowley said. 'He's my mate.'

'I know.' Roulston touched his fingertips to Kitch's wrist, feeling the rapid, thready pulse. 'But face it man, the picture isn't good. Loughnan's very weak. He can't last out the distance.'

'Shut up, you bastard. Stop your bloody bone-pointing,' Rowley snapped. He stood between Roulston and Kitch as if he would protect the sick man from the other doctor's grim prognostication.

'Hey, calm down, doc,' McKone whispered.

Rowley realised that he was close to hysterics and made a determined effort at control.

'Sorry, Roulston. You are right. He can't go on like this for another three or four days. Somehow we have to help him along.'

Rowley struggled to recall everything he had read about pneumonia as a medical student. Astonished, as he had been when he found he could remember Plato's description of the Magnanimous Man, he began to visualise whole sentences of text. In cases where there had been the danger of cardiac failure it had been the practice to induce an artificial crisis. How? Auto-haemotherapy. Blood was drawn from the patient's arm then reinjected into the body, usually the buttock. There was a rapid allergic reaction to the injection of infected blood, a multiplication of leucocytes, a steep rise in temperature as the body mustered its forces to combat the infection.

'We can try auto-haemotherapy,' he told Roulston.

'Lord, man, that's pre-antibiotic Dark Ages. It could work. I'll give you a hand.'

They turned Kitch on his side. McKone swabbed his scrawny buttock while Rowley and Roulston prepared to draw blood from the big vein in the crook of his left arm. Rowley gave the injection.

'We can only wait and see, now,' Roulston said. 'I gather it's no use asking you to get some rest.'

'No.'

For an hour he listened to Kitch's shallow, grunting breathing. Cyanosis was becoming more marked. The lips were blueish although the face was dry and flushed; the nostrils were pinched and the cheeks hollow. The dry eyelids flickered.

'Come on, Kitch, you can make it,' Rowley said aloud.

He became aware that there were patients on the bench at the MIR end of the room. Tommy Moxham was there and so was Stan Manning.

'I must attend to those fellows. Watch Kitch. Call me at any sign of a change,' he told McKone.

Manning had an abscess on his leg. Tommy Moxham complained that he had diarrhoea again and had vomited.

'Okay. You don't have to put on that sad face, Tom. I'll check you over as soon as I've lanced this abscess.'

Rowley swabbed Manning's leg with methylated spirit, Moxham leaning over to watch the operation.

'You're in my light,' Rowley told him. 'And stop breathing on me and Manning.'

He was about to lance the abscess when feet pounded on the wooden floor outside the MIR. Kelk burst in.

'A bloke's just keeled over outside. Looks nasty. One of the British.'

Rowley and Roulston both ran. The man who had collapsed was dead.

'Cardiac beriberi?' Rowley queried.

Roulston nodded. 'Yes, I thought it was only a matter of time.'

McKone bounded out of the building yelling. 'Captain Richards! You're needed . . . Kitch.'

Rowley went back into the hospital faster than he had left it. McKone was at his heels. Kitch was tossing about and throwing back his blankets. His face was beaded with sweat. His cheeks felt cooler to the touch. He was sweating profusely.

'The crisis!'

Rowley realised that he was laughing nervously. He made a determined effort at control.

'Get those wet clothes off Loughnan and keep him wrapped in dry blankets, McKone. Pinch them where you can,' he said. 'Don't you bloody well let him get chilled. I have to lance an abscess.'

When Manning's abscess was lanced and dressed Rowley realised that Moxham was still waiting for attention.

'All right, Tommy, how many days in hospital are we going to give you?' he said.

'I'm not bludging. I feel sick,' Tommy protested.

Rowley looked at him thoughtfully, then looked more closely. Tommy was flushed in the face, especially around the forehead. It looked as if a rash was forming. He was also scratching his scalp.

'Why are you scratching?'

'Ah . . . the bloody lice. Don't give a bloke any peace.'

'Let me see.'

Rowley peered at Moxham's scalp through the bristle of close-cropped hair. There were lice. There were also some small, familiar pimples.

'Now show me your legs.'

There were pimples on Moxham's right shin.

'Oh!'

Rowley felt very weary. He straightened slowly, told Moxham to wait and went out to find Roulston, who had been attending to the removal of the corpse.

'One of my men appears to have developed smallpox,' Rowley told the British doctor.

Roulston, who had also been through a smallpox epidemic on the line, agreed with Rowley's diagnosis. Together they reported to the Japanese medical sergeant and then to Lieutenant Mori, who acted quickly. All working prisoners were recalled immediately and the entire Sakata camp was placed in quarantine. An isolation hospital was established in the small guard house adjacent to the latrine, and Moxham was admitted. Manning, as the most immediate contact, was transferred from the billet into the hospital.

'Well, I'm not complaining. I can rest my leg,' he said cheerfully.

Kitch, cool, lucid and able to drink soup, was also cheerful although weak. 'I won't get it. I've had my close shave,' he prophesied.

When the quarantine arrangements had been completed Rowley and Roulston sat down with the other three British officers for a mug of tea. Peirce, who believed in tea as the universal panacea, had thoughtfully procured a dixieful from the kitchen. 'Well, this has been quite a day,' Rowley said. 'I feel quite tired . . . in fact I feel bloody tired.' He swallowed some tea. 'You know, I don't think I feel very well at all.'

'Actually, old boy, you look flushed,' Roulston observed.

Rowley ran his fingers through the lavatory-brush bristle of his hair which he had had short-cropped only after entering the Sakata camp. He winced. He had touched a sore spot. His fingers explored delicately and found a pimple.

'I'm putting myself in isolation. I have smallpox, too,' he said.

Roulston looked interested. 'You couldn't have caught it from Moxham today. The incubation period is usually seven to twelve days. We must have a carrier in close contact with the hospital. Perhaps the Jap orderly. Haven't you been feeling ill?'

'Oh, I've had a headache and a slight backache but I was too busy with Loughnan to take much notice.'

'You're now going to get that rest you have been needing.'

Next day Rowley's cervical lymph glands were enlarged and painful and his throat was sore. Red blotches appeared on his back and chest and he began having shivering spells. His temperature was 38 degrees centigrade and one vesicle had formed, about a quarter of an inch in diameter and filled with clear fluid. He examined it with interest and noted a haemorrhagic spot in the centre of the base. A petechial or purpuric rash at that stage usually indicated that the disease would be fatal, but this did not trouble him. He had a feeling that he was not going to die for a long time.

Within a few days Manning and one other had developed smallpox. It was in a mild form, uncomfortable until the scabs began to form on the pustules, but tolerable. The eighteen-day quarantine period for the entire camp was a rest that might well be their salvation. Kitch did not develop smallpox and before the eighteen days were up he was pottering about helping McKone nurse the sick. Rowley felt stronger and more optimistic. He enjoyed being bundled up in blankets on the bedding mat, warm and rested for the first time in years, literally. He wished that the quarantine could go on for weeks.

On the seventeenth day Tommy Moxham developed a fresh vesicle. He was applauded by his fellow patients. Roulston, delighted, brought the Jap medical sergeant to view the eruption. The isolation hospital which now contained four dysentery patients as well as the four smallpox ones, was quarantined for another eighteen days. Fifteen days later Moxham again produced a new eruption. He was told to keep up the good work. With luck they would all spend the rest of the war in quarantine.

At the end of the first week in August the rains ceased and the bombers passed over the camp. The prisoners in the isolation hospital crowded to the windows to watch the steady stream of planes. They seemed to be roaring overhead for hours. The wing shapes were clearly visible. The Allied bombing of northern Japan had commenced.

The camp was cut off from all outside news but the prisoners presumed that Sakata was on the flight path for the plastering of towns to the north. On 10 August there was an attack on the shipping in the port of Sakata. From the hospital windows Rowley could see the bombers, engines shrieking, diving on their targets. It occurred to him that the camp could be flattened but he felt no fear even when a small bomb landed so close to the prison compound that one of the British was slightly injured by a flying fragment.

Immediately after that incident Lieutenant Mori gave the prisoners permission to dig slit trenches within the compound. Previous safety precautions had been against fire only, since the Japanese, inhabiting a land plagued by earthquakes and with their aftermath of fire in the ruins of flimsy wood and paper houses, were supremely conscious of and terrified of the ravages of fire. The near miss, however, made Mori appreciate the benefits of slit trenches, and on the whole he was concerned for the welfare of his prisoners.

On 15 August the Japanese staff were assembled for a parade to hear a broadcast speech by the Emperor.

On the following morning the prisoners realised that the Japanese sentries had abandoned the look-out posts. Lieutenant Mori gave orders for work on the trenches to cease without giving any reason. All prisoners were to be innoculated for plague. The Japanese medical sergeant brought supplies for the innoculations but left quickly when Rowley tried to pump him about the reason for the absolute disruption of the camp routine.

'Everyone in the camp is excited,' Roulston reported to the isolation ward. 'We believe the war is over. There hasn't been an official announcement yet but the end can't be more than days away. Even in hours we could be free.'

'Free!'

'We can get out of here. We can go home.'

'We've made it.'

The tiny hospital shook as its inmates jumped and danced.

When they were finally exhausted and happily quiet Tommy Moxham took the opportunity to make an announcement.

'I have some fresh spots on my belly. Two of them,' he said proudly.

The silence was absolute. No one seemed to be breathing. Tommy's fellow inmates stared at him in horror.

'Well, didn't anyone hear. I have . . .'

'Shut up. Tommy,' Stan Manning said.

'But I have two spots on . . .'

'Moxham, you do not have any spots,' Rowley rapped out.

'Yes, I do, doc. Look for . . .'

Rowley glared. 'Moxham, attention!' he yelled. 'Now listen, Moxham, this is an order. You do not have any spots on your belly, you bloody idiot.'

'Moxham, you are expected to obey your officer,' Roulston added austerely.

After evening roll call on 19 August Captain Peirce conducted a Thanksgiving Service which commenced with 'O God, Our Help in Ages Past' and ended with the National Anthem sung with emotion and sincerity as never before. The lessons were read from II Samuel XXII, verses 2–17, and Philippians IV, verses 4–13.

The official announcement was made on 20 August after evening rollcall. Lieutenant Mori addressed the prisoners.

'The longed-for day of Peace has arrived,' he read out. 'You are now no longer prisoners and must obey your own officers, striving for disciplined and restrained behaviour at all times. In this way the relations between yourselves and the civilians of our city should be peaceful. We are concerned for your health and your safety—myself, the guards, and the Kempei [Military Police]—and we shall do our best to protect you should any incidents of ill-feeling occur. You will be given further information when such is available.'

After the address Mori formally handed over the ashes of the deceased prisoners into the safe keeping of Captain Peirce.

The Allied officers, the four British and Rowley, now released from isolation, were then called into the IJA office for a briefing. All POWs would now be under the control of their own officers, who would be responsible for internal discipline, timetables, rollcall and guard duty; there would also be a Japanese guard for the protection of the camp. As there was no guarantee the prisoners would be safe outside the camp, should ill-feeling against the victors sweep through the civilian population, no one was to be allowed beyond the confines of the compound for the time being.

A week of changes, in which the good things that they had dreamed of for years happened, left the prisoners bemused and mentally numb. There was an issue of cigarettes, tinned M & V (meat and vegetables) and tinned oranges. Red Cross medical supplies from Yamagata Military Hospital arrived, as did an issue of Red Cross clothing. The rice ration was increased, on request, from 800 to 1100 grams per man daily. POW signs were painted on the roof of the compound building. A flight of American fighters passed overhead on 25 August, but made no sign of recognition of the POW camp.

Captain Peirce, as the senior British officer in the camp, called a conference of officers. They met in the Orderly Room now furnished with a table and Western style chairs. All of them had regained colour in their faces and their cheeks filled out. Their eyes were alert.

'Chaps, we shall be taking over full control of this camp in less than ten days. We want to do it in style, like the soldiers we are. We should have a full parade for the changing of the guards—show the flag. Any suggestions?'

'I doubt that there are any Allied flags in Sakata,' said Barrett, who had acted as the prisoner's administration officer. 'But I should think there is plenty of red, white and blue silk in this land of the silkworm. We could sew flags.'

'Then we'd better liberate some sewing machines and call for volunteers who can operate them,' Rowley said. 'Between the Union Jack and our stars there'll be a lot of sewing.'

'Right!' Peirce agreed. 'I'll speak to Mori about the sewing machines. We'll have to choose the guard and start drilling so that the ceremony goes without a hitch.'

'Train a number of reserves too,' Roulston advised. 'The men are still in a weakened state. We could have a high rate of collapses on parade.'

'Right. Now I think we should have a small ceremony of our own immediately.'

Peirce delved into a large brown paper bag that he had placed on the table. He produced tiny bowls of thin porcelain and a litre bottle decked in a jacket of finest ricepaper. As he removed the seal he explained, 'Finest number one grade sake as imbibed by the Emperor himself. Liberated from the sake factory across the road by Lieutenant Mori, on request.'

He proposed the Loyal Toast and they drank solemnly. Then they finished the bottle. The sake was the smoothest spirit Rowley had ever tasted.

The billets soon took on the appearance of a backyard rag trade factory. Sewing machines whirred and piles of red, white and blue silk were tumbled on the sleeping decks. When the British began an earnest discussion on the order in which the crosses of the Union Jack were superimposed one on the other, the Australians were amused until Jackie McKone muttered, 'Don't laugh fellers. How many stars do we have in the Southern Cross? And do they lie at an angle or not?' Twenty-four Australians went into an anxious huddle, finally working out that the long axis was vertical and contained five stars, and there was a larger one placed under the Union Jack.

The machinists made stripes for the corporals and sergeants and shoulder tabs for the officers. Rowley insisted on having the red and blue shoulder tabs of the Artillery. The sewing was rough but everyone was properly fitted out.

The next task for the officers was to prepare a nomimal roll of all who had been in the Sakata Camp and a report covering the Japan Party since its formation in Tamarkan in March 1944, the shipwreck of *Rakuyo Maru* and *Kachidoki Maru*, the recovery of survivors, and details of the Sakata Camp from 5 October to the present. Peirce and Pote-Hunt concentrated on the overall report relating to accommodation, rations, working parties and treatment by the Jap guards and civilians. Barrett prepared the nominal roll while Roulston and Rowley wrote the medical report based on their carefully kept medical records.

Rowley also concentrated on writing a detailed diary of his own experiences while the details were still fresh in his memory.

On 26 August Peirce conducted a Memorial Service. The moving and emotional service commenced with 'When I Survey the Wondrous Cross' and after reading the names of the deceased members the service concluded with 'Abide With Me' and the National Anthem. Not one eye was dry as they sang with choked voices. The lesson was from Revelation VII, verses 9 to end.

Pondering over his completed diary and the memories it triggered, Rowley again asked the question: Why did some survive and others not? There was no simple answer. Many factors appeared to play a part. Pure chance and sheer luck could be the answer. Factors such as determination, mateship, improvisation, adjustment of moral values, positive thinking and many others could be attributed to those who survived. But they could also be applied to those who didn't.

Maybe there is a God, after all, who determines our fate—the same God he thought deserted him on some occasions and protected him on others. But why?

The first airdrop by the United States Air Force was made on 28 August. The prisoners gathered in the compound, waving and cheering as a flight of two B17s, A25 and A55, passed overhead. Black objects, bomb-like in appearance, dropped from the belly of A55, but the menacing similarity was dispelled when parachutes opened above the cylinders.

Then one of the B17s buzzed the camp. It came down lower than the top of the chimney of the sake factory. The bomb bay doors were open and the prisoners, jumping, waving and yelling frantically, could see the faces of the crew as they leaned over the safety rail.

A note fluttered down. It was a cheery greeting signed by all the bomber's crew and added that this was their last mission—the thirty-fifth. The message on the note read:

See you Stateside.
This is our last mission (35). Stationed at Saipan.
Address 071 Squadron
497 Bomber Group
San Francisco, California

Passenger	—Chartien
Right Gunner	—Wm. N. Port—Wis.
Left Gunner	—Wm. H. Eichenberger—Pens.
Crew Chief	—Andrew Chappez—Oreg.
Pilot	—Ralph J. Evans—Ind.
Co-Pilot	—Edgar A. Cain—Tex.
Bomber	—Arthur E. Wienshaw—NY
Navigator	—Harold S. Bloomenthal—Vermont
Engineer	—George W. Moore—S. Carolina
Radio Operator	—David H. Weaver—Tex.
R.D.F. Operator	—John K. Breakey—NY
Chief Fire Control	—Virgil E. Furst—Iowa
Tail Gunner	—Theodore B. Ball—Louis.

Allied Prisoners

The Japanese Government has surrendered. You will be evacuated by Allied Nation forces as soon as possible. Until that time your present supplies will be augmented by air drop of U.S. food, clothing and medicines. The first drop of these items will arrive within one hour or two hours.

The drums had come down about a hundred yards south of the compound, causing some damage to Japanese property in the very congested area, for which Peirce later expressed regret. Some drums had come adrift from their parachutes and split on impact, so their contents were ruined, but there were sufficient good things in the undamaged drums to make A55 seem like modern transportation for Santa Claus.

As well as clothing—including shoes in a sufficient range of standard sizes so that almost all of the men, who had gone for so long shoeless, could be fitted—the Yanks had thought to include all the little items that were longed-for luxuries to the prisoners: razors and blades, shaving cream, toothbrushes and paste, combs, toilet soap and sewing kits.

The food cases contained cans of meat, soup, fruit juices, fruit, sugar, milk, cocoa, and real coffee with the almost forgotten luxuries of candy, gum and matches.

There was an orgy of dressing up, shaving and hair-brushing. Rowley, passing from the hospital to the Orderly Room, saw Kitch Loughnan wearing new shirt, trousers and field jacket, shoes and socks, with a handkerchief tucked nonchalantly into the opened breast pocked of the field jacket. He was sitting on a Jap guard's chair in the autumn sunshine, eating a candy bar.

'I look a bit of all right, eh?'

'Your trousers are too short, Kitch.'

'And yours, Captain Richards, sir, might just be a bit too long. What a pity we're all dressed up with nowhere to go. When do you think they'll let us out of here?'

Rowley looked at the Japanese guards who still patrolled the wire.

'After the changeover. Why not? There doesn't seem to be any indication of real ill-feeling among the civilians.'

'No, it's all happened so suddenly. We were the ones who were beaten for so long. Now everything is reversed. They've been conquered. That guard is marching up and down the same as ever, but I wonder what he's feeling.'

Rowley put his hand on Kitch's shoulder. Together they watched the Japanese turn smartly and begin the return on his beat.

The grand change of guards took place on 2 September 1945 at 1200 hours. The final selection of the guard had been made from those considered least likely to collapse. The Australian representative on the guard was Titch Lemin—just turned twenty-one and the youngest Australian. None collapsed. They looked good. As some Dutch and American prisoners were in the camp in transit at the time of capitulation, additional flags had been made. The Japanese flag was lowered and the four Allied flags crept slowly to their mastheads. The Union Jack spread regally in the wind and the Southern Cross was beautiful.

The First Allied Guard consisted of:

Orderly Officer	Captain J.E. Peirce
Orderly Sergeant	Sergeant Hanley
Orderly Corporal	Corporal Balfour
Guards	Driver Smith
	Driver Lemin AIF
	Fusilier Stuart
	Private Peacock
	Gunner Maxted
	Gunner Miller

The office and Japanese sleeping quarters were taken over, and in the afternoon Captain Barrett was in charge of the first outside excursion, a swimming party. That night many of the troops went AWL but there was no trouble in the town. Most of them had been visiting civilians who had been kind to them during the winter. The following evening 225 men were taken to a local theatre and supper. Each man was given a bottle of beer and relations with the civilians were amicable.

Rowley and Kitch dined the next night at Takahashi's house. Their host presented his wife and son and Rowley admired the beautiful kimono that Mrs Takahashi was wearing. Takahashi excused himself and after a short time returned carrying a folded kimono of superb colouring. He offered it to Rowley with outstretched hands.

'Please accept this as a humble gift from my family. It is of a special pattern designed to be worn on happy days. This is a happy day for you.' Takahashi turned his head aside. Quietly he said, 'I think there will be no more happy days in Japan.'

Rowley saw himself again with Kitch, watching the Japanese sentry on patrol. He heard Kitch, at the low table, cough deeply.

'Now we shall drink sake, yes?' Takahashi said.

Stan Manning was on sentry duty at the main gate and he was having trouble. An officer of the Imperial Japanese Army, obviously drunk, was confronting him with drawn sword, cursing in Japanese. Manning lowered his rifle from his shoulder and motioned the Japanese away with his bayonet. 'On your way, mate. You're drunk.'

The Japanese took a step forward, flourishing his sword. He was almost onto the bayonet.

Kitch and Rowley froze. So did a British sentry who had come running to the gate. It was all up to Manning, who reversed his rifle, brought up the butt and cracked the Jap on the chin, knocking him unconscious.

Kitch let out his breath. 'That one will have a sore chin when he wakes up,' he commented.

The British sentry shook his head. 'Cor, how often have we all imagined having a Jap at the end of a bayonet. Why didn't you run the blighter through, Aussie?'

Manning stared at the man sprawled in the gutter and his expression was puzzled. He couldn't imagine himself running a bayonet through anybody in cold blood. It was something you did when you were fighting, but not in cold blood.

He lifted his chin and shouldered arms. 'Because I wouldn't put myself down on the same level as that little yellow bastard,' he said positively.

'Good-night, Sentry.' Rowley said.

'Good-night, Captain.'

'What are you going to do when that fellow comes around?'

'He won't be any trouble. He should be a bit more sober.'

The Japanese officer was already stirring and in a few moments scrambled to his feet. He looked about for his sword, retrieved and sheathed it and walked off into the darkness without a word.

The local mayor hosted a formal dinner for the officers at the best local restaurant where their best china was on display. The officers had learned that it was a Japanese custom that when an item was admired it was presented as a gift. Rowley and the others had no qualms in admiring the saki bottles and cups and accepting them with due formality.

Small groups in the charge of an officer or a senior NCO were free to explore the mysteries of Sakata which had been forbidden to them.

One one occasion when Rowley and his group were looking for trinkets in the shops a Mrs Narita introduced herself. She had spent thirty years in London with her husband in the Japanese legation and was delighted to speak English after some years. With her assistance Rowley purchased a kimono, obi and bag in her special silk shop.

Two days later she entertained Rowley and some friends to a sukiyaki meal at her home. She suggested a visit two days later to the beautifully kept Japanese gardens of a local millionaire and dignitary, Homer. The beauty of the place equalled the sunsets in Burma and the horrors of the past three and a half years just fell away in the atmosphere of peace and tranquillity.

At last on 12 September, exactly twelve months after the shipwreck, the Sakata camp group was alerted to pack and be ready to move out the next morning. The officers liberated all the best quality sake in the factory and each man was issued with a one-litre bottle. Rowley settled down with his little band of Australians in their billet. They began to get very happily drunk, shouting reminiscences and singing. Someone began to improvise: 'Four and twenty black birds baked in a pie. When the pie was opened the birds began to sing. They all sang "Waltzing Matilda" and made the Japs' ears ring.'

It was an instant hit and they all roared out the doggerel lustily, over and over again. Rowley went to the latrines, then decided he should look in on Roulston. The four British officers were seated at a table in the room they had taken over as an orderly room. An American officer had joined them. There were glasses of sake on the table but the officers did not look as if they were celebrating. They looked as if they were holding a wake, Rowley thought. He decided they were suffering from nerves over going back to the outside world again; he had felt that way himself, off and on, during the day. After all, four years had passed and things might not be the same at home.

'Hey, what's the matter with you chaps? Isn't this a celebration?' he said cheerfully.

The American officer looked at him with a faint smile. Roulston, Barrett and Pote-Hunt were observing their sake cups with disapproving expressions. Captain Peirce cleared his throat, 'We don't expect an officer to drink with the men, Captain Richards. It isn't done. You're letting the side down.'

'Well I'll be damned,' said Rowley. 'I'll drink with them because they are my men. And I might get a little drunk with them.'

The next morning he had a hell of a hangover. All his group looked hungover but they had their gear ready and Russell Savage was carrying their flag, properly mounted. The column of British was being assembled. The Australians were already on parade. Warrant Officer Keith Martin formally handed over the parade to Rowley, who proudly led his men marching behind the Australian flag.

Roulston came up and clapped him on the shoulder, then shoved a small bottle into his hand. 'Aspirin all round, old boy. That was quite a celebration.'

Then Roulston was striding away towards the head of the British column.

The ex-prisoners of war marched out of Sakata camp and through the streets of the town to the railway station with their home-made flags flying. Rowley let his men drop back a little from the main body so they marched as a separate group under the Southern Cross. Kitch started the singing of 'Waltzing Matilda'.

By the time they had reached the railway station the euphoria had died away. They commandeered a carriage for their group and slumped into their seats, suddenly very weary. Rowley distributed the aspirin and then they slept while the train rattled on its way across the island to the east coast town of Sendai. When the train finally halted at the station no one was anxious to leave the security of their carriage. There were strangers outside. They waited until an officer looked in and said, in the accents of home, 'Are there any Australians on this train?' He was a naval officer from HMAS *Bataan*. All were suddenly galvanised into action.

'Yes, over here,' somebody called. After a brief look at the Australian group he said quietly, 'I see you've had a tough time'. Rowley was so pent up with emotion that he was unable to speak and through his blurred eyes he saw the officer wipe a tear from his eye.

The Sakata group was embarked on the British destroyer, HMS *Wakeful*, just one year after their voyage on the Japanese frigate, for the trip along the coast to Tokyo Bay. The young officer assigned to them looked after his group of Australians like lost sheep, in fact he positively coddled them. Rowley thought he was rather overdoing it.

'There's no need to treat us like lost sheep,' he said.

The young officer looked startled. 'Well, we want to look after you. You've been through a lot. You are some of the fellows who were on the *Rakuyo Maru*, aren't you? We already have most of your lot from Yokohama. And

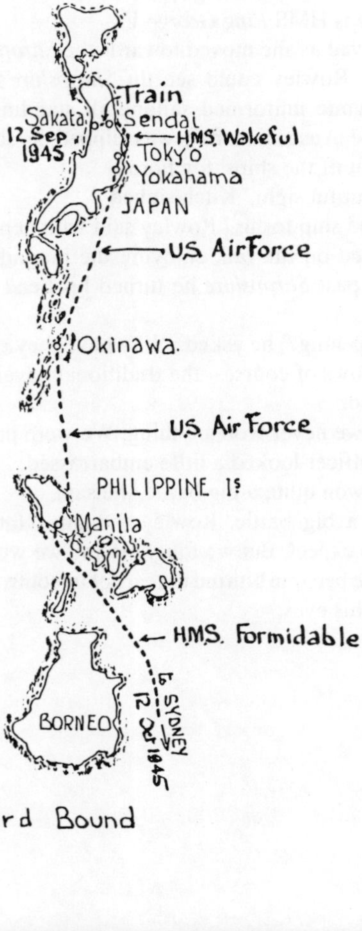

Homeward Bound

. . . you don't know yet, do you? The American submarines picked some of your fellows out of the water. They are already back in Australia.' He hesitated. 'Only eighty altogether, I'm afraid.'

'That's something. We were afraid they'd all been lost,' Rowley said. 'Look, don't worry that we can't take the tough news. Treat us like lost sheep if you like, but not mentally deficient ones. We still have all our marbles.'

The young officer laughed. 'I can see you have. And something else. Sakata was due to be bombed on 17 August. You're luckier than you realise.'

Kitch leaned over and clutched Rowley's arm, 'Look at all the ships anchored along the coast. There are hundreds and hundreds of them. They go on right to the horizon. Cripes!'

'The Allied invasion fleet,' the young officer told them.

No one wanted to comment, only to look and look at all those ships until *Wakeful* ploughed into Tokyo Bay, which was full of ships at anchor, too. There was the giant battleship, USS *Missouri*.

'She took the surrender,' the officer explained. 'And there is the *Shropshire*. Australian. Commissioned in the U.K. on 20 April 1943. Captain J.A. Collins commanding. We'll be passing quite close to her. You can have a good look. And over there is HMS *King George V*.'

Wakeful slowed as she moved towards the *Shropshire* and dropped her Union Jack in salute. Rowley could see tht *Shropshire*'s deck was lined with rows and rows of white uniformed sailors, all standing to attention. Her pennant fluttered as the Australian flag was dipped in salute to the *Wakeful*, to the accompaniment of the ships' fog horns.

'What a beautiful sight,' Kitch sighed.

'She's dressed ship for us,' Rowley said with pent-up emotion.

Russell leaned on the rail, enjoying the magnificent spectacle. As *Wakeful* moved slowly past *Shropshire* he turned his head and queried their attendant officer.

'What's happening?' he asked. 'Why have they dressed ship?'

'For you fellows of course—the traditional naval salute.'

Russell stared.

'For us! But we never won anything. We were prisoners of war.'

The young officer looked a little embarrassed.

'I think you won quite a big battle,' he said.

Yes, it was a big battle, Rowley thought. Not the sort of battle we had been trained to expect. But we fought it and we won.

The spectacle became blurred as tears of uncontrollable emotion and gratitude welled up into his eyes.

Epilogue

Most of the 2/15 Field Regiment medical orderlies returned to Australia to resume their civilian occupations—Jim Armstrong, BEM, to chicken farming near Morriset, N.S.W.; Don Booth to an insurance agent in Yass; Syd Brien to Sydney; 'Pickles' Brawn to the Tramway Department in Sydney; Tom Cooney to Cobar, N.S.W.; Duncan McNaughton as a butcher in Kyogle; 'Pinky' Rhodes as a salesman in Sydney; Allen Taylor to banking in Sydney; Arthur Baker to grazing in Old Junee; Arthur Harris to North Sydney.

Of the twenty-nine in Sakata after the sinking of the *Rakuyo Maru*, twenty-four returned to Australia:

Eddie Anderson became a golf professional in Queensland; Frank Attrill returned to Melbourne and was tragically killed in an accident three days after his return; David Clark returned to farming in Victoria; Ross Dunbar to grazing in Dubbo, N.S.W.; Arthur 'Butch' Hall to business as a butcher in Sydney; Harry Hodgson to farming in Queensland; Teddy Kelk as a clerk in Sydney; Titch Lemin a service station proprietor in Melbourne; Kitch Loughnan to grazing in Mitchell, Queensland; Ian MacDiarmid a successful land developer in Queensland; Jack McKone a driver in the public service and awarded the BEM; Geoffrey Macauley to farming in Victoria; Stan Manning to gold mining in Omeo, Victoria; Keith Martin in business in Sydney; Tom Moxham to grazing in N.S.W.; Alick Robinson a butcher in South Australia; Russell Savage an accountant and company director in Brisbane (he and Ian MacDiarmid married sisters); Jack Smith to bootmaking in Melbourne; brothers Ernie and Jack Wade as commercial travellers in Sydney; Bert Wall a hairdresser in Western Australia; Steve Weeks to driving in N.S.W.; Clarrie Wilson a sign-writer in Brisbane.

The three senior officers lost on the *Rakuyo Maru* on their way to Manchuria were Brigadier Arthur Varley, MC; Group Captain James More, RAF; and Colonel Harry R. Melton, USAF. Others lost on the *Rakuyo Maru* included Jim Burke, John Chalmers, Len Coon, Laurie Phillips, Campbell-Smith, Les Stupart, Arthur Sumner and Bert Tate, DCM.

Many members of 8 Australian Division AIF returned to Australia to achieve distinction in many spheres of life. Lieutenant-Colonel Charles Anderson, VC, MC, returned to grazing in Young, N.S.W. and served two terms in the Federal Parliament; Lieutenant-General H. Gorden Bennett, CB, CMG, DSO, VD, resumed business as a company director in Sydney; Brigadier Cecil Callaghan, CB, CMG, DSO, VD, resumed business in the shoe industry; Sir Albert Coates, OBE, resumed surgical practice in Melbourne and wrote *The Albert Coates Story*, relating his experiences in two wars; Sir Adrian Curlewis, CVO, CBE, became a judge; Colonel Alfred Derham, CBE, MC, ED, resumed medical practice in Melbourne; Brigadier Sir Frederick Galleghan, DSO, OBE, ED, served in Europe post-war and returned to the public service; Captain Ron Greville graduated in medicine and became Secretary of NH&MRC in Canberra; Major-General Don Kerr, CBE, resumed banking in Brisbane and was appointed Civilian Member of the Military Board; Lieutenant John Ross, MC, returned to mining engineering in N.S.W.; Major John Shaw, DSO, became Commissioner of Main Raods in N.S.W.; Lieutenant-Colonel Glyn White, OBE, resumed medical practice in Melbourne and became ADMS Victoria; Roy Whitecross, after graduating in Economics at Sydney University, became an Assistant Registrar of that university and wrote a best-seller *Slaves of the Son of Heaven*, telling of his experiences as a POW; Lieutenant-Colonel John Williams, OBE, ED, resumed engineering, was a witness in the Japanese War Crimes Trials and was active in ex-POW affairs; Lieutenant-Colonel John Wright, DFC, ED, resumed as a company secretary in Sydney and was active in ex-service and community affairs.

Rowley Richards was mentioned in despatches for his service as a POW and later awarded the MBE and ED for his military service in war and peace. He became a Knight of the Most Venerable Order of St John of Jerusalem and was awarded fellowships in Australian Colleges of General Practice and Occupational Health; the Australian Sports Medicine Federation and Institute of Management and the American College of Sports Medicine.

Many of Rowley's friends in the 2/15 Field Regiment and other units returned to Australia. Frank Ball resumed his career as a banker; Rupert Barraclough became a motor company executive; John Bull, who escaped on the *Hong Fat*, was captured in Sumatra and later returned to grazing in N.S.W.; Lloyd Cahill resumed medical practice as an ophthalmologist; Max Carment returned to accountancy; Vic Duncan, BEM, after a further four years in the RAN became a school teacher; Norman Eadie resumed as a consultant surgeon in Melbourne; Ted Fisher as a consultant physician in Sydney; Tom Hamilton, ED, as a surgeon; John Higgin as a physician; Allan Hobbs and Sid Krantz as consultant surgeons in Adelaide; Theo LeMaistre Walker resumed legal practice; John Loewenthal returned to a distinguished career as Dean of the Faculty of Medicine, Sydney University, and was knighted for his services in the Army and to Medicine; Scippy Maher, OBE, became a school headmaster and President of RSSILA, N.S.W.; Des Makepeace resumed accountancy and became general manager of a food chain in Sydney; Fred Martin became Private Secretary to the Governor of N.S.W.; Arthur Moon resumed practice as an obstetrician and gynaecologist in Sydney; Stuart Simpson resumed dental practice in Hobart, Tasmania; Padre Smith returned to the Jesuit Order in Sydney; Jack Varley returned to the family stock and station business in N.S.W.; Stuart Ward resumed business in Sydney; Reg Wilstencroft, MM, returned to the Department of Main Roads in N.S.W.; John Workman graduated post-war as a physiotherapist and practised in Melbourne.

Lieutenant Kimala Nagahashi, captain of the frigate which rescued survivors from the *Rakuyo Maru*, died in action at Leyte in 1944.

After their triumphant entry into Tokyo Bay the Sakata group was transferred from the HMS *Wakeful* to a U.S. hospital ship where they were stripped, deloused in disinfectant, and issued with a full set of new clothing before being flown to Manila via Okinawa. In Manila they made their first real contact with home—Rowley, with others, was permitted to record a radio message to his family and friends to be broadcast in Australia by the Australian Broadcasting Commission.

The Convalescent Hospital in Manila provided far more food than they could eat and all put on weight rapidly—up to 10 kg in some cases. So much for shrunken stomachs!

The Sakata group, together with survivors from other camps in Japan, completed the final leg to Sydney on the aircraft carrier HMS *Formidable*, to be met by cheering crowds of relatives, friends and citizens of Sydney on 13 October 1945.

After minimal formalities all were granted pre-discharge leave. Emotional and exciting homecoming reunions with families and friends were followed by weeks of even more emotional and exciting reunions with fellow ex-POWs to catch up with news of their mates.

Outstanding for Rowley were his reunions with John Shaw and Jim Armstrong. John returned the diary covering the period Christmas Day 1942 to Tamarkan but regretted that he was forced to jettison the bundle of letters home based on the full diary commencing on 7 December 1941, which was summarised on six sheets of paper before leaving Tamarkan and later buried with a graph of sickness rates and rations on Jeep Island.

Jim miraculously returned all of the medical records from Tavoy and the Railway together with a microscope 'acquired' on the way.

Early in 1946 when Rowley resumed work as a Resident Medical Officer at St Vincent's Hospital, Sydney, he spent many nights expanding his diary (the one returned by John Shaw) by talking to Mary Best, a friend from the 1930s who was able to type at speaking speed. This expanded diary covered the period from Christmas Day 1942 to January 1944.

On 15 February 1947 he received from the War Graves Commission the six-page summary, buried under the cross of Corporal Gorlick on Jeep Island, which was recovered several months after Corporal Gorlick's remains had be relocated in the Kranji War Cemetery.

During the immediate post-war years Rowley attended a refresher course in medicine and a postgraduate course in obstetrics and gynaecology. He then spent one year as a research fellow at the Womens' Hospital, Crown Street, in Sydney.

In 1946 Des Makepeace, who had been released in Singapore, married the Colonel's daughter, Barbara Wright, and a year later Rowley married Beth McNab. Each acted as the other's best man and their children grew up as friends.

In 1948 Rowley commenced practice in Seaforth, and with the assistance of John Shaw, who lived in adjacent Balgowlah, designed a house which he built at Balgowlah Heights overlooking Sydney's Middle Harbour, which was the location envisaged for his 'dream home' planned in Burma. The reality, in which he still lives, was considerably more modest!

Ten years later Rowley, accompanied by his wife Beth, had the opportunity to visit Yuzo Matsumoto and Chukichi Takahashi, with whom he had corresponded since his return to Australia. It was a moving reunion which gave Rowley the opportunity to visit familiar places including the billets, the crematorium and Homer's lovely Japanese garden.

Douglas Albion, an executive of *Reader's Digest* in Sydney, learned of the sinking of the *Rakuyo Maru* and urged Rowley to write a book on his experiences as a POW. In 1980 he introduced Rowley to Marcia McEwan, who undertook to assist with a book incorporating Rowley's diaries and medical records. This was after Clay Blair, on a visit to Australia in 1978 to research material for his book *Return from the River Kwai*, had convinced Rowley that he should proceed with a book on his experiences.

In order to refresh and verify his recollections of events and places, Rowley was loaned copies of the diaries of Scippy Maher, Jim Lynch, Roy Whitecross and John Williams to complement his own records. Accompanied by Beth, he then visited Singapore, Malaysia and Thailand to retrace his steps in especially significant places such as Gemas, Muar, Tanglin, Changi, Tamarkan and Kanchanaburi. Later he visited Washington, DC, to study records, particularly relating to the shipwreck and the recovery of survivors, in the National Archives of the U.S. and the U.S. Navy Archives. Some of the photographs taken on these trips are reproduced in this book.

On 22 November 1988 Rowley read the eulogy at the funeral of his lifelong friend, Des Makepeace.

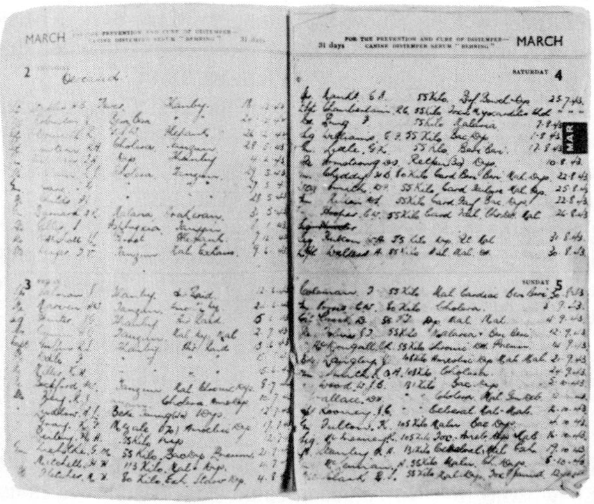

List of deaths in 'A' Force recorded by Jim Armstrong in an old diary.

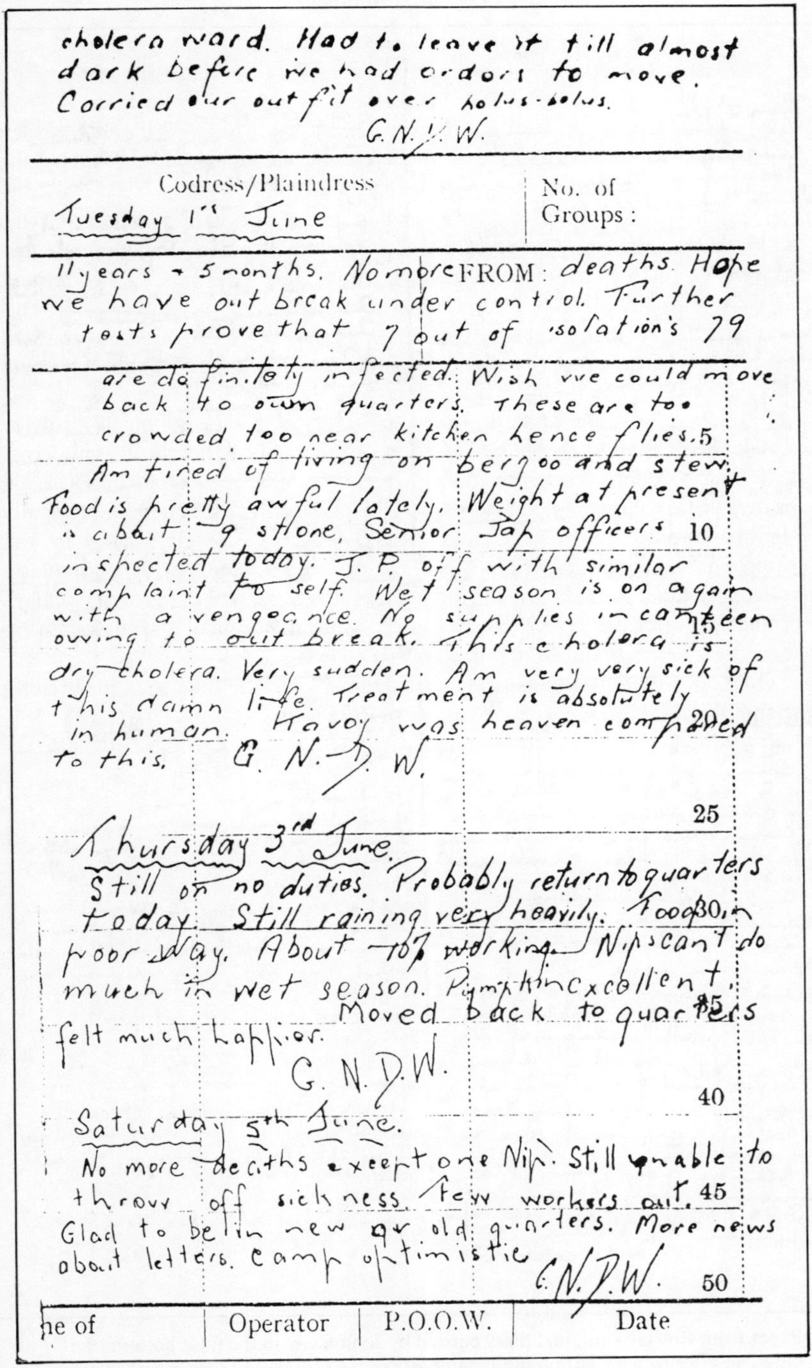
cholera ward. Had to leave it till almost dark before we had orders to move. Carried our outfit over holus-bolus.

G.N.D.W.

Codress/Plaindress | No. of Groups:

Tuesday 1st June

11 years + 5 months. No more deaths. Hope we have out break under control. Further tests prove that 7 out of isolation's 79 are definitely infected. Wish we could move back to own quarters. These are too crowded too near kitchen hence flies. Am tired of living on bergoo and stew. Food is pretty awful lately. Weight at present is about 9 stone. Senior Jap officers inspected today. J.P. off with similar complaint to self. Wet season is on again with a vengeance. No supplies in canteen owing to out break. This cholera is dry cholera. Very sudden. Am very very sick of this damn life. Treatment is absolutely inhuman. Kavoy was heaven compared to this.

G. N. D. W.

Thursday 3rd June.

Still on no duties. Probably return to quarters today. Still raining very heavily. Food in poor way. About 10% working. Nips can't do much in wet season. Pumpkin excellent. Moved back to quarters felt much happier.

G. N. D. W.

Saturday 5th June.

No more deaths except one Nip. Still unable to throw off sickness. Few workers out. Glad to be in new or old quarters. More news about letters. Camp optimistic

G.N.D.W.

ne of | Operator | P.O.O.W. | Date

A page from Scippy Maher's diaries. Courtesy Mrs J. Maher.

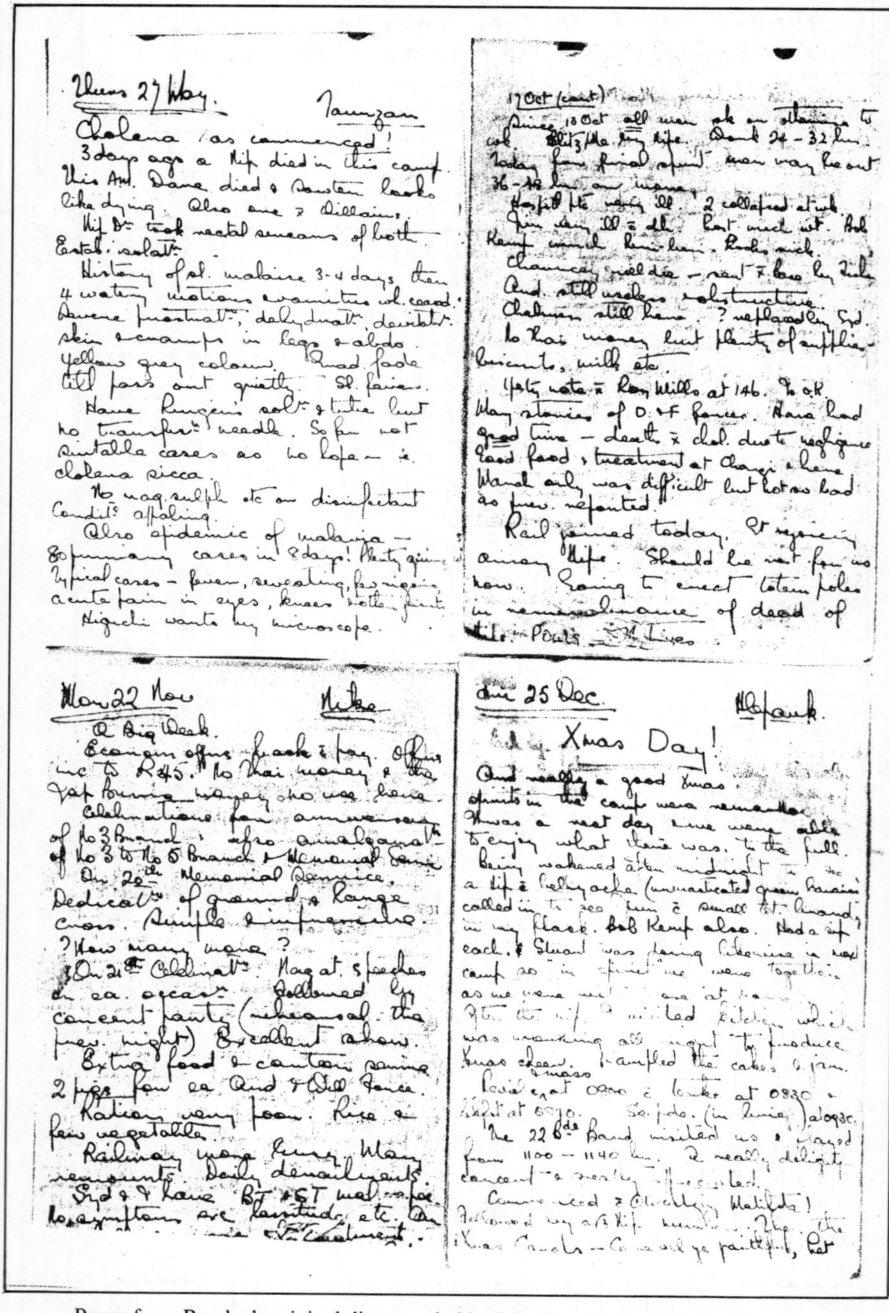

Thurs 27 May. Taungzan

Cholera has commenced!
3 days ago a Nip died in this camp.
This am. Dane died & Sawten looks like dying. Also one [illegible] Dillains.
Nip Dr took rectal smears of both. Estab. isolat[n].
History of sl. malaise 3-4 days then 4 watery motions & vomiting wh. ceased. Severe prostrat[n], dehydrat[n], [illegible] skin & cramps in legs & abdo. Yellow grey colour. Read [illegible] till pass out quietly. Sl. fever.
Have Ringer's sol[n] & tubes but no transfus[n] needle. So far not suitable cases as no hope — i.e. cholera sicca.
No mag. sulph etc or disinfectant. Condit[s] appalling.
Also epidemic of malaria — 80 primary cases in 8 days! Plenty quinine. Typical cases — fever, sweating, [illegible] acute pain in eyes, knees & other joints.
Higuchi wants my microscope.

17 Oct (cont.)
Since 13 Oct all men [illegible] on [illegible] to [illegible]. [illegible] 24 - 32 hrs. Today [illegible] men may be out 36 - 48 hrs on [illegible].
Hospital pts very ill. 2 collapsed at wk.
Jim very ill [illegible] but much wt. Bob Kemp [illegible]. Looks sick.
Chancey [illegible] — [illegible]
And still useless & obstructive.
Chalmers still here. ? replaced by Syd.
No Thai money but plenty of supplies — biscuits, milk etc.
[illegible] Roy Mills at 146. [illegible] O.K.
Many stories of D. & F. forces. Have had good time — death [illegible] chol. due to negligence. Good food & treatment at Changi & here. March only was difficult but not so bad as prev. reported.
Rail joined today. Et rejoicing among Nips. Should be rest for us now. Going to erect totem poles in remembrance of dead of [illegible] Posts [illegible] Lives

Mon 22 Nov Nikhe
A Big Week.
[illegible] offers [illegible]. Offer inc to R45. No Thai money & the Jap Burma [illegible] no use here.
Celebrations for anniversary of No 3 Branch & afso amalgamat[n] of No 3 to No 5 Branch & Memorial Service.
On 20th Memorial Service. Dedicat[n] of ground & large cross. Simple & impressive.
? How many more ?
On 21st Celebrat[n]. Nag at. speeches on ea. occas[n] followed by concert party (rehearsal the prev. night) Excellent show.
Extra food & canteen serving 2 pigs for ea. Aust & [illegible] Force.
Ration very poor. Rice & few vegetables.
Railway more busy. Many [illegible]. Daily derailments.
Syd & I have BT & ST mal. [illegible] no symptoms are lassitude, etc. [illegible] [illegible] treatment.

[illegible] 25 Dec. [illegible]
Xmas Day!
And really a good Xmas. Spirits in the camp were remarkable. It was a rest day & we were able to enjoy what there was to the full.
Being wakened after midnight [illegible] a Nip [illegible] belly ache (unmasticated green bananas) called in to see him [illegible] small [illegible] brandy in my flask. Bob Kemp also. Had a sip each. & Stuart was doing likewise in next camp so "in spirit" we were together as we were [illegible] at [illegible].
After the [illegible] visited kitchen which was working all night to produce Xmas cheer. Sampled the cakes & jam.
[illegible] Xmass at 0800 [illegible] at 0830 [illegible] at 0930. [illegible] (in [illegible]) at 0930.
The 22 Bde Band visited us & played from 1100 — 1140 hrs. A really delightful concert — [illegible] appreciated.
Commenced [illegible] Matildas! followed by a [illegible] Nip [illegible]. The [illegible] Xmas Carols — Come all ye faithful, but

Pages from Rowley's original diary carried by John Shaw in the false bottom of a billy-can after Rowley left on the Japan Party.

Index

1/4
THE
JAPANESE GOVERNMENT
1/4 RUPEE
BC
BC
10
SB
SH
1
THE
JAPANESE GOVERNMENT
M
AD
ONE CENT
M
AD
APANE
VE
10
THE
JAPANESE GOVERNMENT
TEN
CENTS
MX
MX
MM
50
MT
THE
JAPANESE GOVERNMENT
PROMISES TO PAY THE BEARER ON DEMAND
FIFTY CENTS